Lecture Notes in Mathematics

Edited by A. Dold and B. Eckmann

926

Geometric Techniques in Gauge Theories

Proceedings of the Fifth Scheveningen Conference on Differential Equations, The Netherlands
August 23-28, 1981

Edited by R. Martini and E.M.de Jager

Springer-Verlag
Berlin Heidelberg New York 1982

Editors

Rodolfo Martini
Twente University of Technology, Department of Applied Mathematics
Postbus 217, Enschede, The Netherlands

Eduardus M.de Jager
University of Amsterdam, Mathematical Institute
Roetersstraat 15, 1018 WB Amsterdam, The Netherlands

AMS Subject Classifications (1980): 53 C, 58 G, 81 XX

ISBN 3-540-11497-1 Springer-Verlag Berlin Heidelberg New York
ISBN 0-387-11497-1 Springer-Verlag New York Heidelberg Berlin

Printing and binding: Beltz Offsetdruck, Hemsbach/Bergstr.
2141/3140-543210

P R E F A C E

This volume is an account of the lectures delivered at the Fifth Scheveningen Conference on Differential Equations.

The organization of the conference was in the hands of B.L.J. Braaksma (University of Groningen), E.M. de Jager (University of Amsterdam), H. Lemei (Delft University of Technology), R. Martini (Twente University of Technology) and was financially supported by the Minister of Education and Sciences of The Netherlands and Z.W.O. - the mathematical centre.

Like the preceding conferences the meeting was centered around a topic of main interest in differential equations or related fields with the hope of stimulating further research into this area.
This time, as the title indicates, the subject was "gauge theory", in particular Yang-Mills fields, with emphasis on geometric techniques.

The articles included here present a view of gauge theory, embracing many aspects of the subject.
The contributions of R. Hermann and Th. Friedrich are introductions to the geometry of Yang-Mills fields.
The notes of F.A. Bais and P.J.M. Bongaarts deal with the physical background of gauge theories.
E.F. Corrigan's article treats the magnetic monopole, whereas in A. Trautman's paper we find a comparison of Yang-Mills theory and gravitation.
M.G. Eastwood and R.S. Ward lectures are concerned with the twistor approach for dealing with certain "geometric" nonlinear equations of mathematical physics.
Finally, P. Molino reports on prolongation theory, a topic closely related to gauge theory.

It is a pleasure to acknowledge our gratitude to all the authors for accepting our invitation and for their stimulating accounts.
Gratitude is also expressed to Springer-Verlag for their courtesy and efficiency, and we would also like to offer our thanks to Mrs. Lidy Krukerink for her assistance during the period of organizing the conference and in the preparation of the final typescript.

R. Martini
E.M. de Jager
Enschede, Amsterdam, February 1982
The Netherlands.

C O N T E N T S

LIST OF PARTICIPANTS

Invited speakers

F.A. Bais	Institute for Theoretical Physics University of Utrecht
P.J.M. Bongaarts	Lorentz Institute University of Leiden
E.F. Corrigan	Department of Mathematics University of Durham, U.K.
M.G. Eastwood	Mathematical Institute Oxford University, U.K.
Th. Friedrich	Humboldt-University of Berlin, sektion Mathematik, Berlin, D.D.R.
R. Hermann	The Association for Physical and System Mathematics, Brookline, U.S.A.
P. Molino	Institut de mathématiques Université des sciences et Techniques du Langedoc, Montpellier, France
A. Trautman	Institute of Theoretical Physics University of Warsaw, Poland
R.S. Ward	University of Dublin, Trinity College, School of Mathematics, Dublin, Ireland

Other participants

P.J. van Baal	Institute for Theoretical Physics University of Utrecht
M. Bergveld	Institute for Theoretical Physics University of Amsterdam
J.G. Besjes	Department of Mathematics Delft University of Technology
J.H. de Boer	Mathematical Institute University of Nijmegen

B.L.J. Braaksma	Mathematical Institute University of Groningen
P.M. van den Broek	Department of Mathematics Twente University of Technology
R.H. Cushman	Mathematical Institute University of Utrecht
D. Dunnebier	Department of Mathematics Delft University of Technology
B. Dijkhuis	Mathematical Centre, Amsterdam
H. ten Eikelder	Department of Physics Eindhoven University of Technology
W.T. van Est	Mathematical Institute University of Amsterdam
J.F. Frankena	Department of Mathematics Twente University of Technology
J.A. van Gelderen	Department of Mathematics Delft University of Technology
J. de Graaf	Department of Mathematics Eindhoven University of Technology
P.K.H. Gragert	Department of Mathematics Twente University of Technology
E.W.C. van Groesen	Mathematical Institute University of Nijmegen
G.J. Heckman	Institute for Theoretical Physics University of Utrecht
E.M. de Jager	Mathematical Institute University of Amsterdam
P. Jonker	Department of Mathematics Twente University of Technology
E.A. de Kerf	Institute for Theoretical Physics University of Amsterdam

P.H.M. Kersten	Department of Mathematics Twente University of Technology
G.W.M. Kallenberg	Department of Mathematics Delft University of Technology
T.H. Koornwinder	Mathematical Centre, Amsterdam
H. Lemei	Department of Mathematics Delft University of Technology
R. Martini	Department of Mathematics Twente University of Technology
W.M. de Muynck	Department of Physics Eindhoven University of Technology
H.G.J. Pijls	Mathematical Institute University of Amsterdam
J.W. de Roever	Department of Mathematics Twente University of Technology
E. Sitters	Mathematical Institute University of Amsterdam
H. Rijnks	Department of Mathematics Delft University of Technology
G.M. Tuynman	Mathematical Institute University of Amsterdam
F. Twilt	Department of Mathematics Twente University of Technology
B.J. Verhaar	Department of Physics Eindhoven University of Technology

FIBER SPACES, CONNECTIONS AND YANG-MILLS FIELDS

Robert Hermann

The Association for Physical and System Mathematics

Research supported by Ames Research Center (NASA), NSG-2402; U.S. Army Research Office, #LILG1102RHN7-05 MATH; and The National Science Foundation MCS 8003227.

ABSTRACT

From the point of view of a differential geometer, Yang-Mills Fields are connections on principal fiber bundles whose curvature satisfies certain first-order differential equations. These lecture notes assume a knowledge of the formalism of calculus on manifolds, i.e., the theory of differential forms and vector fields, and are based on the theory of connections in fiber spaces, developed primarily by E. Cartan and C. Ehresmann in the period 1920-1955. To make the material more readily accessible to someone familiar with classical physics, the emphasis will be on Maxwell electromagnetic theory, considered as a Yang-Mills with an *abelian* structure group. Some of the material is from *Interdisciplinary Mathematics*, some is new.

TABLE OF CONTENTS

1. INTRODUCTION

There has been a long and glorious tradition of mutually energizing and enriching interchange between differential geometry and mathematical physics. Let us refer only to Gauss, Riemann, and Einstein! However, in the recent past this interchange almost ceased: physicists could not understand the post-tensor analysis differential-geometric formalism, based on the coordinate-free, quasi-algebraic methods of "calculus on manifolds," and mathematicians, under the intellectual influence of topology, emphasized purely *global* and non-computational problems. (Physicists need to be able to calculate and think about geometric formalisms in more intuitive ways.) The typical developments of the 1950's and 1960's in differential geometry were the Index Theorem and the Rauch Comparison Theorem, both magnificent achievements, but unintelligible (at least in the short term) to physicists. Of course, with the development of more global aspects of gauge field and nonlinear wave theories (for example, to explain the "confinement" of quarks), this sort of mathematics is becoming much more interesting to them.

My lectures will not be about these global aspects of modern geometry, but will attempt to show how rich are the possibilities for application of the underlying *Cartan-Ehresmann* methodology. Of course, I cannot make any serious historical analysis here. Let me only say that the roots of this methodology lie in the period 1860-1920, in the work of such mathematicians as Riemann, Darboux, Poincare, Klein, Lie, Goursat, Cartan himself in his early work, Fuchs, Picard, Vessiot, Schlesinger, Ricci, Bianchi, Hilbert, Caratheodory and Levi-Civita. For cultural and intellectual reasons (for example, World War I and the decimation of a generation, the rise of a new axiomatic-set theoretic paradigm in mathematics), this branch of mathematics went into eclipse. It was certainly very difficult to understand in any case, since it involved a mixture of algebraic, analytic, and topological tools that were then only very primitively developed, and it was cursed by abominable notation and language. However, it did produce one mighty tree: Elie Cartan's work in the period 1915-1950. We now look to his *Collected Works* as the essential transition between this lost historical tradition and the present-day science.

I add Charles Ehresmann to this pantheon because his work is an essential link between us and Cartan. His thesis on the topology and geometry of Grassmann manifolds [16] is one of the major documents in the development of the topological side of modern geometry, a synthesis of the ideas emerging from the work by Cartan and de Rham on the role of differential forms in topology and the work of Lefschetz on intersection theory. It turned out to be basic to the theory of *characteristic classes* [78]. He was (along with Whitney, Hopf, and Steenrod) among the founders of fiber bundle theory. It was he who most strongly emphasized the importance for *differential geometry* of the fiber bundle formalism. His work with Georges Reeb on foliation theory [81] is basic to much of present day differential topology, and contains many ideas that have not been fully developed and exploited to this day. His theory of "jets" of mappings is a formalism that has proved very useful in many areas of modern geometry and analysis. But above all, we remember him for one paper [17], which in amazingly few lines sets up a formalism--*the theory of connections*--with which we could understand much of that magnificent geometric tradition cited above. What is even more amazing was that this is precisely the correct geometric formalism for three branches of present day science:

The theory of Maxwell-Yang-Mills and gauge fields of elementary particle physics,

The theory of solitons and nonlinear waves, and associated areas in traditional partial differential equation-oriented applied mathematics and "nonlinear physics,"

The theory of "isomonodromy deformation" and the links to statistical mechanics [72].

Now, what Ehresmann did was to lay down the path toward a rational study of the geometric objects called "connections in fiber spaces," in the same way as groups, fields, rings, etc. had been studied in 20th century algebra. Most of the examples he had in mind were the connections associated with Riemannian metrics and homogeneous spaces of Lie groups. The parts of mathematical physics cited above do not involve only these geometric structures, but others which are defined by the physical differential equations. The prototype in physics is the *Kaluza-Klein theory* [66] of electromagnetism and gravitation. For a differential geometer, the key observation is that the connections and bundles are not determined by purely geometric considerations, but by analytic conditions, differential equations which, locally, involve the components of the connections. Of course, it is no surprise to the geometer to learn that these equations involve the curvature! Thus, in the Kaluza-Klein theory, the gravitational field is determined by a (pseudo) Riemannian metric on a four dimensional manifold, the electromagnetic field by a fiber bundle with one-dimensional fiber over this manifold. The curvature of this metric is then a vector valued two-differential form, the curvature of the line bundle a two-differential form (essentially just the electromagnetic field) with values in the fiber of the bundles. The "Einstein-Maxwell equations" are then certain relations connecting these curvature tensors and the Hodge divergence operator of the curvature tensor of the bundle. Alternately, the complete set of equations can be replaced by a set of conditions on the Ricci tensor of a five dimensional Riemannian metric for the total space of the bundle.

Looked at geometrically, the Yang-Mills fields are a natural generalization of this theory, with the line bundles replaced by vector bundles with non-abelian structure groups. The possibility of the algebraic structure of the Lie algebra of these groups is left open, to give physicists many possibilities of phenomenological analysis and "grand unification."

In my book in the Interdisciplinary series [66] titled *Yang-Mills, Kaluza-Klein and the Einstein Program* I described Yang-Mills theory from this point of view. My aim in this paper is to develop the Yang-Mills theory from the original Ehresmann point of view. (The pioneer work in this approach has been that of Trautman [88].) Given a fiber space

$$\pi: X \to Z,$$

a connection (in Ehresmann's sense) can be defined as a linear map

$$h: \mathcal{V}(X) \to \mathcal{V}(Z)$$

from vector fields on X to vector fields on Z, satisfying certain conditions. This translates some geometrico-physical concepts into algebraic ones. I believe that this is a useful formalism for treating in a more unified way many of the ideas of present day elementary particle-nonlinear waves and non-linear physical theory.

When I was invited to give these lectures, my plan was to emphasize the historical continuity with Maxwell theory. In fact, there are many interesting geometric aspects of electromagnetic theory from this point of view that have never been adequately appreciated or developed. (I have in mind, particularly, work of Gabriel Kron and K. Kondo on the geometric structure of electro-mechanical systems [76,77], as well as many areas in mainstream theoretical physics, such as the theory of monopoles and Feynman path integrals for electromagnetic fields.) However, faced with the sheer quantity of information to convey, I have retreated from these ambitious ideas to the mathematician's more traditional approach of laying down the general principles of the theory mainly in non-historical and non-inductive terms.

However, there is also a positive side to this applied neo-Bourbakiism. I believe (and, of course, have tried to show in my books) that there is a unified geometric methodology that is useful across a wide spectrum of contemporary science and engineering.

Both present-day differential geometry and its 19th century precursors are not "pure" subjects with their own distinctive methods and specialties, but are *hybrids* of topology, algebra, and analysis. Of course, this is nothing new; think of Descartes and "Analytic Geometry," which translates geometric questions into problems of algebra and analysis. (This goes both ways: There have recently been great successes in a variety of important analytic and algebraic problems by means of translation into geometric terms.) Thus, anyone writing in a broad manner about present-day differential geometry (pure or applied) must make a choice about the level of algebraic and analytic expertise, notation, and "language" to be emphasized. My own personal choice is this: Analysis at the "calculus-on-manifolds" level; algebra at the level of van der Waerden, with some additional knowledge of such specialized topics as Lie algebra theory and multilinear-tensor algebra; topology at the level of rudimentary point set topology-manifold theory. I prefer to try to do without systematic use of such undeniably useful and important topics as sheaf and scheme theory, general commutative algebra and nonzero characteristic algebraic geometry and "global analysis." Thus, I position myself somewhere in between what a physicist or applied mathematician might reasonably be expected to know (or intuit) and what an expert in one of the branches of modern geometry considers minimal knowledge *for his specialty*.

Finally, this program is in two parts: Develop the theory of the geometric structures involved in present-day gauge-field-nonlinear wave theory (i.e., certain types of connections in certain types of fiber bundles), then develop their *deformation* theory. (However, due to problems of space, there will not be much explicitly stated about the deformation theory.) Note the terms isomonodromy and isospectral *deformation* have already entered mathematical physics. However, the *geometric* deformation theory involved there is never made explicit. I plan to show (in another place) that they both involve deformation theory for *connections*.

Perhaps a few words are appropriate about the historical background of geometric deformation theory. The prototype is Riemann's theory of "modulii" of Riemann surfaces, i.e., their classification under complex analytic diffeomorphisms. Kodaira amd Spencer [73] generalized and developed Riemann's ideas in such a way that it became clear in the 1950's that the appropriate setting was a deformation theory of the G-structures (in the sense of Cartan, Ehresmann, and Chern). This can be understood in the following way:

A geometric structure on a manifold M is determined as a cross-section of a fiber bundle $E \to M$, with a structure group K. A cross-section is thus a map $\gamma: M \to E$. Let Γ denote the set of geometrically relevant cross-sections.

An equivalence relation must be defined on Γ: Two cross-sections are equivalent if they determine the same "geometric structure." Let Γ' denote the space of equivalence classes. The object of *geometric deformation theory* is to parameterize these points of Γ', for example, by identifying Γ' with the space of cross-sections of a fiber space.

The Riemann theory is the prototype. Let M be a compact two-dimensional orientable manifold. A *complex analytic structure* on M is determined by a cross-section of a certain fiber bundle, with structure group $GL(2,R)$. The fiber of this bundle is $GL(2,R)/GL(1,\mathbb{C})$. (The "G" of the G-structure is the group which occurs when the fiber is a coset space L/G.) A cross-section $\gamma: M \to E$ then determines a complex analytic structure for M, i.e., makes it into a *Riemann surface*. (In this case--accidentally--there are no differential equations required for γ, since an almost complex structure for a two-dimensional manifold is *automatically* complex. In higher dimensions this is not so, of course.) Two such γ's are *equivalent* if there is a diffeomorphism of M into itself which carries one of the complex structures into the other--the modulii "problem"--which is not completely solved to this day--is to describe these equivalence classes.

As in all of geometry, these geometric questions can be translated into *algebraic* and *analytic* ones. The algebraic framework, particularly of deformation theory, is more tractable and attractive. It has led to new and interesting algebraic problems, such as the deformation theory of Lie and associative algebras, which were treated by Gerstenhaber, Nijenhuis, and Richardson. (As a side point to physicists, it led to the first appearance of "graded" Lie algebras.) This theory is also of great physical interest, particularly when it is combined with Lie group harmonic analysis.

There is another theme from my work that I want to mention here--the theory of what I call *gauge* and *current* Lie algebras [43, 46, 42, 38]. Let Z be a manifold and let $\mathcal{F}(Z)$ be the commutative associative algebra of all C^{∞}, real-valued functions on Z. Let $\mathcal{K}$ be the Lie algebras over the real numbers. Consider the tensor product

$$\mathcal{K} \otimes \mathcal{F}(Z)$$

as a module over $\mathcal{F}(Z)$.

One can define various Lie algebra structures on this module that are combinations of the Lie algebra bracket over $\mathcal{K}$ and the differential operators on $\mathcal{F}(Z)$. The simplest one is just the tensor product of the Lie structure in $\mathcal{K}$ and the associative algebra structure on $\mathcal{F}(Z)$. This (following the physical motivation) is called the *gauge Lie algebra.* It will appear in this paper in relation to the theory of connections and gauge fields. Others which typically are deformations of this one, also occur in the physics literature (for example, in what physicists call the σ- or Sugawara model) and are called *current algebras*. The gauge group can be considered as the infinite Lie group of maps $Z \to G$ with the product of two such maps just the point-wise product. More general groups of this sort (leading to "current algebras" as their Lie algebra) may be considered the Ehresmann jet-spaces of maps of Z into G, $J^r(Z,G)$. For more details of this relation between "current algebras" and Ehresmann jets, see "Interdisciplinary Mathematics," Volume 6. That there are relations between this algebra and the theory of connections (hence to "gauge fields") is more-or-less geometrically obvious. See *Lie Algebras and Quantum Mechanics* and *Fourier Analysis on Groups and Partical Wave Analysis* for an early version of this.

I will begin this exposition of the methodology with the theory of what Cartan called Pfaffian Systems [5, 11].

2. PFAFFIAN SYSTEMS

The notation of calculus and manifolds are those of [67]. All manifolds, maps, and geometric data will be C^{∞}, finite dimensional and paracompact unless mentioned otherwise. Let X be such a manifold. Here is the notation we shall use for the basic objects of the calculus of manifolds used in this work.

$\mathcal{F}(X)$ = algebra of C^{∞}, real-valued functions on X.

For $x \in X$, X_x = tangent vector space to X at x.

X_x^d = dual space to X_x, the space of one-covectors at x.

$T(X) = \{(x,v): x \in X, v \in X_x\}$, the tangent vector bundle.

$T^d(X) = \{(x,\theta): x \in X, \theta \in X_x^d\}$, the cotangent bundle.

$\mathcal{V}(X) = C^\infty$ cross-sections of $T(X)$, the derivations of (X).

$\mathcal{D}^1(X)$ = cross sections of $T^d(X)$, the dual (X)-module to (X).

$\mathcal{D}^n(x)$ = exterior differential forms of degree n.

$d: \mathcal{D}^n(X) \to \mathcal{D}^{n+1}(X)$, *exterior derivative*

$(V_1, V_2) \to [V_1, V_2] = V_1 V_2 - V_2 V_1$, an R-linear map.

$[\ ,\ ]: \mathcal{V}(X) \times \mathcal{V}(X) \to \mathcal{V}(X)$, called *Jacobi bracket*. (It is also called *Lie bracket*: Lie himself called it "Jacobi!")

If θ is a differential form and V is a vector field, then

$\mathcal{L}_V(\theta)$ denotes the *Lie derivative* of θ by V.

$G^m(X_x)$ denotes the Grassmann manifold of m-dimensional linear subspaces of the vector space X_x, $m = 0,1,2,\ldots$

$G^m(T(X))$ denotes the fiber bundle over X whose fiber over $x \in X$ is $G^m(X_x)$.

$G^m(T^d(X))$ denotes the bundle of m-dimensional linear subspaces of the cotangent vector spaces.

If $E \to X$ is a fiber space over X, let $\Gamma(E)$ denote the space of C^∞ cross-section maps: $X \to E$.

<u>Notational Remark</u>: The standard mathematical notation for the dual space to a vector space is V*. This often conflicts with notation in mathematical physics, where "*" often means "complex conjugate" or "Hemitian conjugate of operators," hence I have improvised a compromise notation: "V^d."

<u>Definition</u>: A *Pfaffian system* (of dimension m) on X is a cross-section map

$\gamma: X \to G^m(T(X))$, (i.e., an element of $\Gamma(G^m(T(X))$).

Given such a cross-section $\gamma \in \Gamma(G^m(T(X)))$, one can define the "dual" object,

$\gamma^d \in \Gamma(G^{n-m}(T^d(X)))$

$\gamma^d(x)$ = space of $\theta \in X_x^d$ such that $\theta(\gamma(x)) = 0$, i.e., the annihilator of $\gamma(x)$ in the dual space.

γ^d is actually the object which Cartan would call a "Pfaffian system," since he preferred to work with differential forms rather than vector fields. However, we will call either γ or γ^d a "Pfaffian system." (When considering such systems *with singularities*, it is necessary to distinguish one from its dual.)

3. THE DERIVED SYSTEMS OF A REGULAR PFAFFIAN SYSTEM

Keep the notation of Section 2. Let $\gamma \in \Gamma(C^n(T(M)))$ be a Pfaffian system interpreted geometrically as a cross-section of the Grassmann bundle of m-dimensional linear subspaces of the tangent vector bundle to X. Set:

$$\mathcal{V}(\gamma) = \{V \in \mathcal{V}(X): V(x) \in \gamma(x), \text{ for all } x \in X\} \tag{3.1}$$

The orbit curves of the Pfaffian system are the curves in X that are orbit curves of *some* vector field on $\mathcal{V}(\gamma)$. (Note that $\mathcal{V}(\gamma)$ is not a Lie subalgebra of $\mathcal{V}(X)$, unless the Pfaffian system is completely integrable in the Frobenius sense, i.e., defines a foliation.) Set:

$$\mathcal{V}^1(\gamma) = \mathcal{V}(\gamma) + [\mathcal{V}(\gamma), \mathcal{V}(\gamma)] \tag{3.2}$$

$$\gamma^1(x) = \mathcal{V}^1(\gamma)(x), \text{ for } x \in X \tag{3.3}$$

Thus, $\gamma^1(x)$ is a linear subspace of the tangent space that contains $\gamma(x)$. As x varies, we get a family $x \to \gamma^1(x)$ of tangent spaces. Let us say that γ is *one-regular* if the dimension of these spaces is *constant*. In this case, γ^1 defines another Pfaffian system, which was called by Cartan the *first derived system*. Similarly, set:

$$\mathcal{V}^2(\gamma) = \mathcal{V}(\gamma) + [\mathcal{V}(\gamma), \mathcal{V}(\gamma)] + [\mathcal{V}(\gamma), [\mathcal{V}(\gamma), \mathcal{V}(\gamma)]] \tag{3.4}$$

$$\gamma^2(x) = \mathcal{V}^2(\gamma)(x) \tag{3.5}$$

Let us say that γ is *two-regular* if it is one-regular and if dim $\gamma^2(x)$ is constant as x ranges over X. In this case, γ^2 defines a Pfaffian system called the *second derived system*. Continue in this way to define the *n-th derived system*. The original system γ is said to be *regular* if all its derived systems are regular. In this paper, we shall consider only regular systems. (One can prove that, in general, there is always an open subset of X on which the system is regular. If the system is real analytic, this subset is also dense in X, and its complement is contained in analytic varieties of lower dimension.)

In many of Cartan's papers of Pfaffian systems, the study of the properties of the derived system is the key feature. We now turn to the study of the *structure tensors*, which live on the vector bundles associated with the derived systems.

4. THE STRUCTURE TENSOR OF A REGULAR PFAFFIAN SYSTEM

Continue with γ as a regular Pfaffian system on a manifold X, considered as a cross-section of the Grassmann bundle. Let $\gamma^1, \gamma^2, \ldots$ denote the derived systems. For $x \in X$, we have an increasing sequence of tangent subspaces:

$$\gamma(x) \quad \gamma^1(x) \quad \gamma^2(x) \quad \ldots$$

Then $\gamma^1, \gamma^2, \ldots$ also define vector bundles over X. We will make no notational distinction between the γ's as cross-sections of the Grassmann bundles and as vector bundles over X. Set:

$$E^1(x) = \gamma^1(x)/\gamma(x); \quad E^2(x) = \gamma^2(x)/\gamma^1(x) \; ; \; \ldots \tag{4.1}$$

Each $E^1(x)$, $E^2(x), \ldots$ is a linear vector space. As x varies, they define *vector bundles* $E^1, E^2, \ldots$ over X. They play a basic role in the study of the structure of Pfaffian systems.

We shall now define *structure tensors*, $\tau^1, \tau^2, \ldots$ tensor fields associated with the vector bundles defined by $\gamma^1, \gamma^2, \ldots$ and $E^1, E^2, \ldots$ To define τ^1, again work at a point x of X. Pick $v_1, v_2 \in \mathcal{V}(\gamma)$. Then, V_1, V_2 (x) is a tangent vector to X, which lies in $\gamma^1(x)$. Consider its projection mod $\gamma(x)$, i.e., as a vector in $E^1(x)$. We obtain a skew-symmetric map

$$(V_1, V_2) \to \tau^1(V_1, V_2).$$

Note now that τ only depends on the values of V_1 and V_2 at x, not on their derivatives. τ thus defines a skew-symmetric, bilinear map

$$\tau^1(x) : \gamma^1(x) \times \gamma^1(x) \to E^1(x) \quad .$$

Explicitly,

$$\tau^1(x)(V_1(x), V_2(x)) \equiv [V_1, V_2](x) \bmod \gamma(x) \quad . \tag{4.2}$$

As x varies, we obtain a bilinear vector bundle map

$$\tau^1 : \gamma \times \gamma \to E^1 \quad . \tag{4.3}$$

This is called the *first integrability tensor*. Notice that, by the very definition of E^1, τ is onto, hence it is zero if and only if the Pfaffian system with which we began is Frobenius integrable.

We can now continue. For $V_1, V_2, V_3 \in \mathcal{V}(\gamma)$, consider the triple commutator: $[V_1, [V_2, V_3]](x)$. It lies in $\gamma^2(x)$. Its nontensorial component can be eliminated by projecting mod $\gamma^1(x)$. We thus obtain a trilinear map:

$$\tau^2(x) : \gamma(x) \times \gamma(x) \times \gamma(x) \to \gamma^1(x)/\gamma(x) \equiv E^2(x).$$

Now, as x varies, we obtain a tensor field τ^2 as a trilinear bundle map $\gamma \times \gamma \times \gamma \to E^2$. This procedure can obviously be itereated to obtain n-th degree *structure tensors* $\tau^n : \gamma \times \ldots \times \gamma \to E^n$. We will now discuss in what sense these tensors are "invariants" for the equivalence problem.

5. THE STRUCTURE TENSORS OF PFAFFIAN SYSTEMS AS INVARIANTS FOR THE EQUIVALENCE PROBLEM

Let

$$\gamma: X \to G^m(T(X)) \text{ and } \gamma': X' \to G^m(T(X'))$$

define Pfaffian systems (of the same dimension) on manifolds X and X'.

Definition. γ and γ' are *strongly equivalent* (in the sense of Lie and Cartan) if there is a diffeomorphism $\phi: X \to X'$ such that

$$\gamma'(\phi(x)) = \phi_*(\gamma(x)) \text{ for all } x \in X. \tag{5.1}$$

In words, the natural actions of ϕ on the tangent bundle and on the associated Grassmann bundles intertwine the cross-section maps γ and γ'.

Remark. Of course, this general definition is not how Cartan and Lie would think of it! Cartan, for example, would deal locally with "moving frames," i.e., locally defined column vectors of independent one-forms X and X':

$$\omega = \begin{pmatrix} \omega_1 \\ \vdots \\ \omega_{n-m} \end{pmatrix} \tag{5.2}$$

$$\omega' = \begin{pmatrix} \omega'_1 \\ \vdots \\ \omega'_{n-m} \end{pmatrix} \tag{5.3}$$

such that, for $x \in X$, $x' \in X'$,

$$\gamma(x) = \{v \in X_x : \omega(v) = 0\} \tag{5.4}$$

$$\gamma'(x') = \{v' \in X'_x : \omega'(v') = 0\} . \tag{5.5}$$

Relation (5.1) now means that there is a (n-m) × (n-m) matrix M of functions on M such that:

$$\omega = M\phi^*(\omega') \quad . \tag{5.6}$$

A "weaker" concept of equivalence involves *prolongation* of the two systems, then transformation of one system into another by a differomorphism. In terms of the theory of differential equations, this would provide a way of allowing solution of differential equations to coincide without the spaces in which the solutions live being diffeomorphic.

Return to relation (5.1). If ϕ satisfies (5.1), it also satisfies

$$\phi_*(\mathcal{V}(\gamma')) = \mathcal{V}(\gamma') \quad . \tag{5.7}$$

From (5.7) it follows that ϕ acts on the derived systems $\gamma^1, \gamma^2, \ldots$, the associated vector bundles, and intertwines the action of the structure tensors. The "algebraic invariants" of these structure tensors will be "equivalence invariants" of the Pfaffian systems.

6. SOLUTION MANIFOLDS AND LINEAR VARIATIONAL EQUATIONS OF PFAFFIAN AND EXTERIOR DIFFERENTIAL SYSTEMS

It is a major 19th century idea (emphasized most strongly by Poincaré) to study nonlinear differential equations (particularly stability properties) by linearization about a "nominal" solution. Of course, this is still a vitally alive idea in both pure and applied mathematics! Since a variant of it plays a strong role in both nonlinear wave theory and isomonodromy deformation theory, I will review certain general principles.

Let $\gamma: X \to G^m(T(X))$ be a Pfaffian system, considered as a cross-section of a Grassmann bundle. Let Y be another manifold and let $\phi: Y \to X$ be a submanifold map.

Definition. ϕ is a *solution* of the Pfaffian system if the following condition is satisfied:

$$\phi_*(Y_y) \subset \gamma(\phi(y)) \text{ for all } y \in Y \quad . \tag{6.1}$$

Remark. In local coordinates for Y and X, one sees that condition (6.1) defines a set of partial differential equations. However, it is often counterproductive to think of them in this way: their great virtue is that they are "geometric," i.e., independent of local coordinates. Cartan convincingly demonstrates in [9] that this is a useful way of thinking about differential equations, particularly those which come from geometric problems. Even for those equations coming from physical situations, it is often useful to translate them into Cartan's form, particularly in the search for their *symmetries* [12,13].

Conditions (6.1) are often more conveniently viewed in terms of differential forms. An *exterior differential system* $\mathcal{E}$ on X is a collection of differential forms which is an ideal in the Grassmann algebra, and such that:

$$d\mathcal{E} \subset \mathcal{E} \quad .$$

A Pfaffian system $\gamma: X \to G^m(T(X))$ defines an exterior differential system $\mathcal{E}(\gamma)$. Consider the one-differential forms θ on X such that:

$$\theta(\gamma(x)) = 0 \text{ for all } x \in X \quad . \tag{6.2}$$

$\mathcal{E}(\gamma)$ is then defined as the smallest differential ideal of forms containing these dual one-forms.

Definition. A submanifold map $\phi: Y \to X$ is a *solution* of the exterior differential system $\mathcal{E}$ if the following condition is satisfied:

$$\phi^*(\theta) = 0 \text{ for all } \theta \in \mathcal{E} \, . \tag{6.3}$$

Remark. Dieudonne's treatise [15] is the most accessible treatment in the modern literature of the Cartan-Kahler existence theorem for exterior differential systems [14].

If $\phi: Y \to X$ is such a solution manifold, one can define a set of linear differential equations defined along the submanifold $\phi(Y)$. Let

$$t \to \phi_t \quad , \quad 0 \le t \le 1$$

be a one-parameter family of submanifold mappings. For $y \in Y$, let

$$v(y) = \text{tangent vector to the curve } t \to \phi_t(y) \text{ at } t = 0$$

$\underline{v}$: $y \to v(y)$ is a map $Y \to T(X)$. Then, if θ is a differential form on X,

$$\frac{\partial}{\partial t}\phi_t^*(\theta)\Big|_{t=0}(y) = \phi^*(v \lrcorner d\theta) + d\phi^*(\underline{v} \lrcorner \theta) \tag{6.4}$$

Thus, if ϕ_t is a solution submanifold for each t (or to "first order in t"), then $\underline{v}$ satisfies:

$$\phi^*(\underline{v} \lrcorner d\theta) + d\phi^*(\underline{v} \lrcorner \theta) = 0 \quad \text{for all } \theta \in \mathscr{E} . \tag{6.5}$$

These are (as linear differential equations for $\underline{v}$) the *linear variational equations* of the exterior system $\mathscr{E}$ and the solution ϕ.

One can also look at this from the Lie point of view. If V is a vector field on X, then, for $\theta \in \mathscr{E}$,

$$\mathscr{L}_V(\theta) = d(V \lrcorner \theta) + V \lrcorner d\theta . \tag{6.6}$$

To solve the linear variational equations means to look for vector fields V such that:

$$\phi^*(\mathscr{L}_V(\theta)) = 0 \tag{6.7}$$

or

$$d\phi^*(V \lrcorner \theta) + \phi^*(V \lrcorner d\theta) \tag{6.8}$$

Let $N(\phi)$ be the *normal vector bundle* to the submanifold N. It is defined as the real ordered pairs

$$(y, \underline{y}) , \tag{6.9}$$

where $y \in Y$, and $\underline{y}$ is an element of the vector space

$$X_{\phi(y)}/\phi_*(Y_y) , \tag{6.10}$$

which is, geometrically, the space of normal tangent vectors to the submanifold $\phi(Y)$

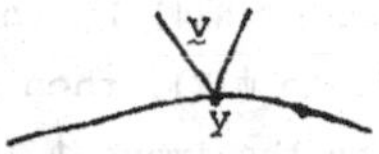

(As usual in linear algebra when there is no quadratic form available to identify an "orthogonal complement" to a linear subspace of a vector space, the "normal vectors" are identified with the *quotient* vector space.)

It is readily verified that these equations only depend on the values V taken on the submanifold $\phi(Y)$ and are the projection of V in the normal bundle $N(\phi)$. Thus, the equations (6.5) determine a set of *linear* differential equations for cross-sections of the vector bundle $N(\phi)$.

7. FOLIATIONS, THE EHRESMANN-REEB HOLOMONY GROUP, AND THE BOTT CONNECTION

Let $\gamma: X \to G^m(T(X))$ be a Pfaffian system. γ is said to define a *foliation* for X if the first structure tensor vanishes. This means that

$$[\mathcal{V}(\gamma), \mathcal{V}(\gamma)] \in \mathcal{V}(\gamma) \quad , \tag{7.1}$$

i.e., $\mathcal{V}(\gamma)$ is a Lie subalgebra of $\mathcal{V}(X)$.

The Frobenius Complete Integrability Theorem (in its modern "global" version) asserts that, for an γ satisfying (7.1) and for each $x \in X$, there is a unique maximal connected solution submanifold of X of dimension m, called the *leaf* of the foliation through x.

Let $\phi: Y \to X$ be such a leaf. Let $V \in \mathcal{V}(X)$ and $\theta \in \mathcal{D}^1(X)$ be a vector field and a Pfaffian form on X such that

$$\theta(\mathcal{V}(\gamma)) = 0 \quad , \tag{7.2}$$

i.e., θ belongs to the dual Pfaffian system. Then, $\phi^*(\theta) = 0$. The linear variational equations (6.8) are then:

$$\phi^*(V \lrcorner d\theta) + d\phi^*(V \lrcorner \theta) = 0 \tag{7.3}$$

But, for $W \in \mathcal{V}(\gamma)$,

$$(V \lrcorner d\theta)(W) = d\theta(V,W) = V(\theta(W)) - W(\theta(V)) - \theta([V,W]) = -W(\theta(V)) - \theta([V,W]) \tag{7.4}$$

<u>Theorem 7.1</u> There is a flat linear connection on the normal vector bundle $N(\phi)$ to a leaf $\phi: Y \to X$ of a folation γ on the manifold X. The flat cross-sections of the connection are then the solutions of the linear variational equations of the underlying Pfaffian system.

<u>Proof</u>. One obtains cross-sections of $N(\phi)$ by the following geometric process: Take vector fields V on X, restrict it to $\phi(Y)$, then project it to the quotient vector space of the tangent bundle T(X) by the image $\phi_*(T(Y))$ of the tangent bundle to Y, i.e., the tangent vectors to the submanifold $\phi(Y)$. Denote this cross-section by $\underset{\sim}{v}$. Let W be a vector field tangent to $\phi(Y)$. Set:

$$\nabla_{\underset{\sim}{W}} v = \text{projection of } [W,V] \text{ in the normal bundle} \tag{7.5}$$

It is readily verified that this operator defines a bona fide linear connection in $N(\phi)$. It is readily seen that the flatness of the connection, in other words, the vanishing of the curvature tensor, follows from the fact that $\mathcal{V}(\gamma)$ forms a Lie subgebra of $\mathcal{V}(X)$, i.e., that the Pfaffian system γ is "completely integrable."

As for any linear connection, one can define the *holomony group* of this connection on $N(\phi)$. It is a group of linear transformations on the fiber of $N(\phi)$ above one fixed point. Since the connection is flat, the holomony group is discrete. It is the *obstruction* to the leaf space of the foliation being a manifold, in the following sense:

<u>Theorem 7.2</u>. If there is a manifold Z and a submersion map $\pi: X \to Z$ whose fibers are the leaves of the foliation, then the holomony group of the foliation is the identity.

If the holomony group of the foliation is not the identity, it plays a role in the study of the *stability* properties of the leaves, i.e., in the sense in which leaves which come close at one point remain (or do not) live at other points.

Remark. Surprisingly (especially in view of his role as the founder of the theory of connections in fiber bundles!) Ehresmann did not use this natural definition of the holomony group of a foliation. Instead, he was guided by his more general notion of "pseudogroups," and R. Bott provided [4] this definition, and proved interesting consequences.

8. FIBER SPACES AND THEIR "GAUGE" LIE ALGEBRAS

We now move on to the central topic of this paper. Let X and Z be manifolds. Let

$$\dim X = n, \quad \dim Z = m \quad .$$

Let $\pi: X \to Z$ be a *submersion map* from X to Z, i.e., the map $\pi_*: T(X) \to T(Z)$ on tangent vectors is *onto*. Thus, by the implicit function theorem, X can *locally* be parameterized by:

$$(y,z), \quad y \in R^{n-m}, \quad z \in R^m,$$

with

$$\pi(y,z) = z \quad . \tag{8.1}$$

The geometric object (X,Z,π) will be called a *fiber space*.

For each $z \in Z$, the inverse image

$$\pi^{-1}(z) \equiv X(z)$$

(called the *fiber* above z) is, by the implicit function theorem, a submanifold of X. (It is, in fact, a regularly imbedded submanifold, i.e., its intrinsic topology as a manfiold is the topology induced as a subset of X.) The tangent vectors fo X(z) at $x \in X(z)$, are the vectors $v \in T(X)$ such that: $\pi_*(v) = 0$. These vectors are called the *vertical vectors*.

A vector field $V \in \mathcal{V}(X)$ is said to be *projectable* under π if there is a vector field V' on Z such that

$$\pi_*(V(x)) = V'(\pi(x)) \text{ for all } x \in X \quad . \tag{8.2}$$

Alternate geometric ways of putting this condition are important for many geometric and physical applications. Here is another one.

$$\pi^*(\mathcal{L}_{V'}(\theta)) = \mathcal{L}_V(\pi^*(\theta)) \text{ for } \theta \text{ a differential form on } Z.$$

If V generates a one-parameter group $t \to \exp(tV)$, i.e., if V is a complete vector field of diffeomorphisms of X, for each $t \in R$, the diffeomorphism $\exp(tV)$ maps each fiber of π into another fiber.

If (y,z) is a local coordinate system fo X satisfying (8.1), then V and V' are of the following form:

$$V = a(y,z)\frac{\partial}{\partial y} + b(z)\frac{\partial}{\partial z} \quad ,$$

$$V' = b(z)\frac{\partial}{\partial z} \quad .$$

Let us state formally some properties of the projectable vector fields that play a role in later sections. (The proofs are routine.)

Theorem 8.1. For each vector field $V \in \mathcal{V}(X)$, there is at most one vector field $V' \in \mathcal{V}(Z')$ such that $\pi_*(V) = V'$.

Let $\mathcal{G}(\pi)$ be the set of all these projectable pairs (V,V'), i.e., $\mathcal{V}(\pi)$ is a subset of the direct sum Lie algebra $\mathcal{V}(X) + \mathcal{V}(Z)$. Then:

a) $\mathcal{G}(\pi)$ is a Lie subalgebra of $\mathcal{V}(X) + \mathcal{V}(Z)$

b) The map $V \to \pi_*(V)$ is a Lie algebra homomorphism of $\mathcal{G}(\pi)$ onto a Lie subalgebra of $\mathcal{V}(Z)$. The map

$$(V,V') \to V$$

is a Lie algebra isomorphism of $\mathcal{G}(\pi)$ with a Lie subalgebra of $\mathcal{V}(X)$. This Lie subalgebra of $\mathcal{V}(X)$ is called the Lie algebra of vector fields which are *projectable* under π.

The Lie algebra $\mathcal{G}(\pi)$ is the basic algebraic structure associated with a fiber space. These vector fields are also central to what the physicists call the theory of *gauge fields*. Accordingly, I will call $\mathcal{G}(\pi)$ the *gauge* (Lie) *algebra* of the fiber space.

I will now try to explain some of the mixed geometric-physical intuition in back of this central concept.

A pair (ϕ,α) of diffeomorphisms

$$\phi: X \to X, \quad \alpha: Z \to Z$$

is called a *gauge transformation* of the fiber space (X,Z,π) (alternate mathematical term: *fiber space automorphism*) if the following condition is satisfied: $\pi\phi = \alpha\pi$. Consider (ϕ,α) as an element of the direct product group: $\mathrm{DIFF}(X) \times \mathrm{DIFF}(Z)$.

In fact, this defines the set of all gauge transformations as a *subgroup*, denoted as $G(\pi)$, of: $\mathrm{DIFF}(X) \times \mathrm{DIFF}(Z)$. (It is defined *locally* by means of differential equations, hence is an "Infinite Lie Group" in the sense of Lie and Cartan.)

Let y be a point of z, and let $Z(z) = \pi^{-1}(z)$ be the fiber of the fiber space π above the point z. Let $(\phi,\alpha) \in G(\pi)$. Then,

$$\pi\phi(X(z)) = \alpha\pi(X(z)) = \alpha z \quad ,$$

whence:

$$\phi(X(z)) = X(\alpha(z)) \quad . \tag{8.3}$$

(8.3) means that ϕ maps fibers of π (parameterized by points of Z) onto fibers. The action of α tells us how ϕ affects the parameters of these fibers. Now, it is readily seen that the infinitesimal generator vector fields of one-parameter subgroups of $G(\pi)$

are the vector fields in $\mathcal{G}(\pi)$. Thus, $\mathcal{G}(\pi)$ is, in a sense, the "Lie algebra" of the "Lie group" $G(\pi)$. (Unfortunately, the relation between "Lie groups" and "Lie algebras" is not as tight for the "infinite dimensional" ones as for the "finite dimensional." Rather than go into the technicalities of functional analysis needed to make this connection more precise, I prefer to follow the physicists and keep things loose!)

Now, the fiber spaces that occur naturally in different parts of physics come with special geometric structures on the fibers $\{X(z): z \to Z\}$ of π. Thus, it is natural to focus attention on the *subgroup* of $G(\pi)$ consisting of the transformations which preserve this structure.

For example, a common situation [47] is that where the fibers are vector spaces, i.e., π is a *vector bundle*. One is then interested in the subgroup of the (ϕ,α) such that $\phi: X(z) \to X(z)$ is, for each $z \in Z$, a *linear* automorphism of the vector space. Such transformations are then *automorphisms* of the vector bundle, and play a basic role in elementary particle physics, Lie group representation theory (via the theory of "induced representations"), inverse scattering structures, etc. Another interesting possibility (e.g., in relativity-gravitational theory and the various "non-linear" models of elementary particle theory) is that where each $Z(z)$ is a Riemannian manifold and the map ϕ preserves this Riemannian structure.

9. CONNECTIONS IN FIBER SPACES

Continue with $\pi: X \to Z$ as a fiber space (i.e., submersion) map. For each $x \in X$, let

$$VE(x) = \{v \in X: \pi_*(v) = 0\}$$

be the *vertical vectors* at x, i.e., the vector which are tangent at x to the fibers of π. The assignment $x \to VE(x)$ defines a (completely integrable) Pfaffian system; VE $\mathcal{V}(X)$ denotes the cross-sections. VE has two sorts of algebraic structures:

a) VE is a Lie subalgebra of $\mathcal{V}(X)$

b) VE is an $\mathcal{F}(X)$-module.

Of course, the module and Lie algebra structure are not algebraically compatible, since VE is not a Lie algebra over the ring $\mathcal{F}(X)$. However, it has a Lie algebra over the subring $\pi^*(\mathcal{F}(Z))$ of $\mathcal{F}(X)$. To prove this, note that:

$$[V_1,fV_2] = V_1(f)V_2 + f[V_1,V_2] \text{ for } V_1,V_2 \in VE,\ f \in \mathcal{F}(X) \quad .$$

The first term on the right hand side prevents the Jacobi bracket from defining a Lie algebra over the ring $\mathcal{F}(X)$. However, if $f \in \pi^*(\mathcal{F}(Z))$, then $V_1(f) = 0$, since V_1 is tangent to the fibers of π, and f is constant on the fibers. Hence, we have

$$[V_1,fV_2] = f[V_1,V_2] \text{ for } f \in \pi^*(\mathcal{F}(Z)),\ V_1,V_2 \in VE \quad ,$$

which is precisely the condition that VE is a Lie module *over the ring* $\pi^*(\mathcal{F}(Z))$.

It will turn out that this structure of *Lie algebra-over-the-ring* $\pi^*(\mathcal{F}(Z))$ is basic to both the theory of gauge fields and the theory of connections.

<u>Definition</u>. A *connection* for the fiber space π is a Pfaffian system

$$x \to H(x) \qquad X_x$$

on X which is *complementary* to VE, i.e., for each $x \in X$, X_x is the direct sum of the oinear subspaces $VE(x)$ and $H(x)$. Given such a connection, let $\mathcal{H}$ denote the cross-sections of the vector bundle $x \to H(x)$. $\mathcal{H}$ is then an $\mathcal{F}(X)$-submodule of $\mathcal{V}(X)$. Then, each vector field $V \in \mathcal{V}(X)$ can be decomposed (by linear algebra) into a sum $V' + V''$ where, for each $x \in X$:

$$V'(x) \in VE(x), \quad V''(x) \in H(x) \quad .$$

The possibility of such a decomposition is, *for fixed* x, just a matter of linear algebra. However, it is important to realize that V' and V" *depend in a* C^∞ *way on* x, i.e., themselves define vector fields. Then, the *module* $\mathcal{V}(X)$ is a *direct sum* of submodules:

$$\mathcal{V}(X) = VE \oplus \mathcal{H} \quad .$$

This is the algebraic viewpoint. It should be supplemented with the geometric. Let z be a point of Z, and let $x \in X(z)$ be a point in the fiber above z. Think of X as lying above Z, as follows:

$$\begin{array}{ccc} X & & \cdot\, x \\ \downarrow & & \downarrow \pi \\ Z & & \cdot\, z \end{array}$$

Think of the fibers X(x) as curves in X.

The tangent vectors, i.e., elements of VE(x), are tangent to these curves

H(x) may be thought of as the space of values of $\mathcal{H}$ at x. It is a complementary subspace to VE(x).

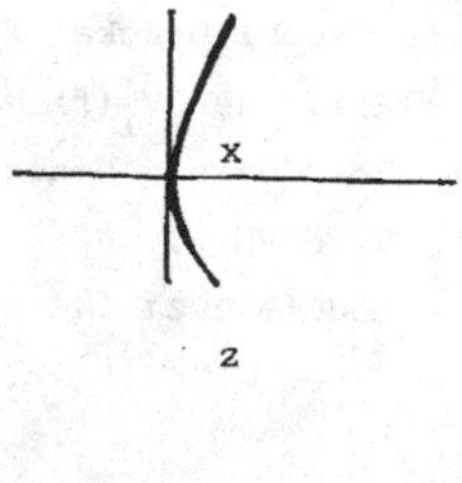

Since $VE(x)$ is the kernel of π_*, i.e., the fiber of π "to the first order," π_* maps vectors in $H(x)$ in a one-one way. Ehresmann called them a set of *horizontal vectors* at x (whence, the notation H and $\mathcal{H}$). Thus, as x varies, we obtain a family of tangent subspaces to X that represents the geometry of the base Z "to first order." This is Ehresmann's intuition for what Cartan meant by "the method of the moving frame."

Consider such an $\mathcal{H}$ as fixed. Let us bring in the gauge algebras $\mathcal{G}(\pi)$, i.e., the set of pairs (V,V') of vector fields, $V \in \mathcal{V}(X)$, $V' \in \mathcal{V}(Z)$, such that $\pi_*(V(x)) = V'(\pi(x))$ for all $x \in X$. For this section, it is convenient to identify (V,V') with V, i.e., to identify $\mathcal{G}(\pi)$ with a Lie subalgebra of $\mathcal{V}(X)$. Set:

$$\mathcal{G}(\pi,\mathcal{H}) = \mathcal{G}(\pi) \cap \mathcal{H} \tag{9.1}$$

<u>Theorem 9.1</u>. Consider π_* as a Lie algebra homomorphism of $\mathcal{G}(\pi) \to \mathcal{V}(Z)$. The kernel is then VE, the Lie algebra of vertical vector fields. π_* is then one-one on the linear subspace $\mathcal{G}(\pi,\mathcal{H})$. $\mathcal{G}(\pi)$ is a direct sum (as a vector space) of VE and the linear subspace $\mathcal{G}(\pi,\mathcal{H})$.

<u>Proof</u>. Since $H(x) \cap VE(x) = (0)$, for all x, we have: $\mathcal{H} \cap VE = (0)$. We have already proved that $\mathcal{V}(X)$ is a direct sum (as a vector space) of VE and $\mathcal{H}$. That $\mathcal{G}(\pi) = VE + \mathcal{G}(\pi,\mathcal{H})$ follows.

<u>Theorem 9.2</u>. π_*, considered as a linear map: $\mathcal{G}(\pi\ \mathcal{H}) \to \mathcal{V}(Z)$ is an isomorphism.

<u>Proof</u>. We have just seen that it is one-one. To prove that it is onto, let $V' \in \mathcal{V}(Z)$. One can construct a cross-section $V: x \to V(x)$ of the vector bundle $x \to H(x)$ of horizontal tangent vectors as follows:

> For $x \in X$, $V(x)$ is the unique element of $H(x)$ which goes into $V'(\pi(x))$ under the isomorphism $\pi_*: H(x) \to Z_{\rho(x)}$.

To finish the proof, we have only to prove that V is smooth, i.e., defines a (C^∞) vector field on X. This follows from the implicit function theorem (since π is a submersion).

Let

$$h: \mathcal{V}(Z) \to \mathcal{H} \quad \mathcal{V}(X) \tag{9.2}$$

be the inverse of π_* restricted to $\mathcal{G}(\pi,\mathcal{H})$. The connection is clearly determined by h.

<u>Theorem 9.3</u>. Consider both $\mathcal{V}(Z)$ and $\mathcal{G}(\pi)$ as $\mathcal{F}(Z)$-modules. Then, h is $\mathcal{F}(Z)$-linear.

<u>Proof</u>. The action of $\mathcal{F}(Z)$ on $\mathcal{V}(Z)$ is just the natural multiplication of vector fields on Z by functions on Z

$$(f,V) \to fV: z \to f(z)V(z) \quad . \tag{9.3}$$

Let $\mathcal{F}(Z)$ act on $\mathcal{G}(\pi)$ as follows:

> Embedd $\mathcal{F}(Z)$ on $\mathcal{F}(X)$ as the subring $\pi_*(\mathcal{F}(Z))$. Then, let $\pi^*(\mathcal{F}(Z))$ act on $\mathcal{G}(\pi)$ by multiplication, as in (9.7), i.e., $fV = \pi^*(f)V$.

That h is $\mathcal{F}(Z)$-linear is now evident from the fact that $\mathcal{F}(Z)$ is identified with functions on X which are constant on the fibers.

10. HORIZONTAL COMPLETENESS OF CONNECTIONS

Keep the notation of Section 9. Consider a connection as defined by an $\mathcal{F}(Z)$-linear, one-one map $h: \mathcal{V}(Z) \to \mathcal{G}(\pi)$. Recall that a vector field V on a manifold M is said to be *complete* if it is the infinitesimal generator of a one-parameter group of diffeomorphisms on M. We can then denote the one parameter group as $t \to \exp(tV)$, $-\infty < t < \infty$. Then, if θ is a differential form on M, we have:

$$\frac{\partial}{\partial t} \exp(tV)^*(\theta) = \mathcal{L}_V \exp(tV)^*(\theta) \tag{10.1}$$

In the real analytic case, exp(tV) is given by *Lie series*:

$$\exp(tV)^*(\theta) = \theta + t\mathcal{L}_V(\theta) + \frac{t^2}{2!}\mathcal{L}_V^2(\theta) + \dots \tag{10.2}$$

Definition. The connection h is said to be *horizontally complete* if the following condition is satisfied:

For each horizontally complete vector field V on Z, h(V) is also horizontally complete.

Remark. Ehresmann included "horizontal completeness" in his definition of "connection."

The horizontal completeness can be used to prove some useful global properties of the fiber spaces that carry them. I shall state a few of them without proof here.

Theorem 10.1. If the fiber space $\pi: X \to Z$ admits a horizontally complete connection, then it is a "local product" fiber space, in the sense that each point $z \in Z$ has an open neighborhood U that the fiber space $\pi^{-1}(U)$ above U is isomorphic to the product fiber space

$$U \times \pi^{-1}(z) \quad .$$

Theorem 10.2. Suppose that X and Z are manifolds of the same dimension. Let $\pi: X \to Z$ be a fiber space map, which, in this case, is just a map with everywhere nonzero Jacobian, i.e., a local diffeomorphism. There is only one connection, namely: $\mathcal{H} = (0)$. It is then horizontally complete if and only if π is a covering map.

11. CURVATURE

Let us recapitulate the notation. X and Z are manifolds, $\pi: X \to Z$ is a fiber space map. $\mathcal{G}(\pi)$ is the Lie subalgebra of $\mathcal{V}(X)$ consisting of the vector fields which are projectable under π. $\mathcal{G}(\pi)$ is an $\mathcal{F}(Z)$-module (but not an $\mathcal{F}(Z)$-Lie algebra). The vertical vector fields VE form a Lie ideal and an $\mathcal{F}(Z)$-submodule of $\mathcal{G}(\pi)$ (as well as an $\mathcal{F}(Z)$-Lie algebra). A *connection* will be defined as an $\mathcal{F}(Z)$-linear mapping

$$h: \mathcal{V}(Z) \to \mathcal{G}(\pi) \quad ,$$

such that

$$\mathcal{G}(\pi) \text{ is a direct sum (as an } \mathcal{F}(Z)\text{-modeul of VE and } h(\mathcal{V}(Z)) . \tag{11.1}$$

$$\pi_*(h(V)) = V \quad , \quad \text{for } V \in \mathcal{V}(Z) \quad . \tag{11.2}$$

Now, h is a map from one Lie algebra to another. However, it is not necessarily a Lie algebra homomorphism. We can construct an algebraic object which measures the deviation of h from a Lie algebra homomorphism.

$$\Omega(V_1,V_2) = [h(V_1),h(V_2)] - h([V_1,V_2]) \tag{11.3}$$

<u>Theorem 11.1</u>. Ω, defined by formula (11.2) is an $\mathcal{F}(Z)$-bilinear map: $\mathcal{V}(Z) \times \mathcal{V}(Z) \to$ VE.

<u>Proof</u>. First, let us prove that the vector field on the right hand side of (11.2) belongs to E, i.e., is tangent to fibers of π. Recall that all vector fields in $\mathcal{G}(\pi)$ are projectable under π, and that the map $V \to \pi_* V$ on projectable vector fields preserves Jacobi brackets. Hence,

$$\pi_*(\Omega(V_1,V_2)) = \text{ , using (11.2) and the right hand side of (11.3),}$$
$$[\pi_*(h(V_1)), \pi_* h(V_2)] - [V_1,V_2]$$
$$= 0 \quad , \quad \text{whence: } \Omega(V_1,V_2) \in \text{VE} \quad .$$

It is obvious from formula (11.3) that $V_1,V_2 \to \Omega(V_1,V_2)$ is R-bilinear and skew-symmetric. We must prove that it is $\mathcal{F}(Z)$-bilinear. Let $f \in \mathcal{F}(Z)$. Then,

$$\begin{aligned}
\Omega(fV_1,V_2) &= [h(fV_1),h(V_2)] - h([fV_1,V_2]) \\
&= [\pi^*(f)h(V_1),h(V_2)] - h(f[V_1,V_2] - V_2(f)V_1) \\
&= \pi^*(f)[h(V_1),h(V_2)] - h(V_2)(\pi^*(f))h(V_1) \\
&\quad - \pi^*(f)h[V_1,V_2] - \pi^*(V_2(f))h(V_1) \\
&= \pi^*(f)([h(V_1),h(V_2)] - h([V_1,V_2])) + 0 \\
&= \pi^*(f)\Omega(V_1,V_2) \quad .
\end{aligned}$$

This is what one means by $\mathcal{F}(Z)$-bilinearity, since $\mathcal{F}(Z)$ acts on $\mathcal{G}(\pi)$ by multiplication by elements of $\pi^*(\mathcal{F}(Z))$.

The $\mathcal{F}(Z)$-bilinearity of Ω, as expressed by Theorem 11.1 indicates its tensorial nature. It is called the *curvature tensor*.

Now, h determines a non-singular Pfaffian system H on X.

$$H(x) = h(\mathcal{V}(Z))(x) \quad \text{for } x \in X \quad . \tag{11.4}$$

Elements of H(x) are called *horizontal vectors*.

<u>Theorem 11.2.</u> The curvature tensor Ω vanishes if and only if the Pfaffian system H defined by (11.4) is completely integrable.

<u>Proof.</u> Ω zero implies (using 11.3)) that $h(\mathcal{V}(Z))$ is a Lie subalgebra of $\mathcal{V}(X)$, whence, using (11.4), that H is completely integrable.

Conversely, suppose that H is completely integrable. Let

$$\mathcal{H} = \{V \in \mathcal{V}(X) : V(x) \in (H(x)), \text{ for } x \in X\} \quad .$$

Then, "complete integrability" means that

$$[\mathcal{H},\mathcal{H}] \quad \mathcal{H} \tag{11.5}$$

We can then (using the Frobenius Theorem) find a local coordinate system for X consisting of functions

$$z, y^a$$

$$1 \le i, j \le m = \dim Z$$

$$m + 1 \le a,b \le n = \dim X \quad ,$$

and associated "flat" vector fields:

$$\partial_i \equiv \frac{\partial}{\partial z^i} \quad , \quad \partial_a = \frac{\partial}{\partial y^a}$$

such that:

$$dz^i(\partial_a) = 0 \quad , \quad dy^a(\partial_i) = 0 \quad , \quad dz^i(\partial_j) = \delta^i_j \quad , \quad dy^a(\partial_b) = \delta^a_b$$

The following conditions are satisfied:

z^i are constants on the fibers of π, i.e., are the pull-back of a coordinate system on the base Z

$$dy^a(\mathcal{H}) = 0 \quad . \tag{11.6}$$

Let V,V' be vector fields on Z. Then h(V), h(V') can be written in terms of these coordinates:

$$h(V) = A^i\partial_i + A^a\partial_a \quad , \qquad h(V') = B^i\partial_i + B^a\partial_a \quad .$$

The definition of h as a connection requires that:

$$\pi_*(h(V)) = V \quad , \quad \pi_*(h(V')) = V \quad . \tag{11.7}$$

(11.7) implies that

$$h(V)(z^i) = A^i \quad \text{and} \quad h(V')z^i) = B^i$$

are the pull-backs of functions on Z. These conditions require that

$$\partial_a(A^i) = 0 = \partial_a(B^i)$$

(11.6) requires that: $A^a = 0 = B^b$. Hence

$$h(V) = \pi^*(dz^i(V))\partial_i \quad \text{for each vector field V on Z.} \tag{11.8}$$

(To make sense of this formula, the dz^i should be interpreted as forms on Z.) Thus,

$$\Omega(V,V') = h([V,V']) - [h(V), h(V')]$$
$$= h([A^i\partial_i, B^j\partial_j]) - [A^i\partial_i, B^j\partial_j]$$

But, using (11.8), we see that:

$$h([A^i\partial_i, B^j\partial_j]) = [A^i\partial_i, B^j\partial_j] \quad ,$$

which proves that $\Omega(V_1,V_2) = 0$.

Since V_1, V_2 are arbitrary vector fields on Z, it follows that $\Omega = 0$, as required to finish the proof.

Locally, we see that curvature zero connections are not very interesting. However, from the global point of view, this is not at all so, especially for physics; for example, both the "isomonodromy" and "isospectral" deformations that occur in soliton-nonlinear wave theory involve curvature zero connections. The following result (of Ehresmann) is one of the simplest global theorems.

Theorem 11.3. Let $\pi: X \to Z$ be a fiber space with a horizontally complete connection. The horizontal vector fields $\mathcal{H}$ then define a foliation of X. Let $\phi: Z' \to X$ be a leaf of this foliation. Then, the composite map

$$\pi\phi: Z' \to Z$$

is a covering map. In particular, if Z is simply connected, then $\pi\phi$ is a diffeomorphism, and the fiber space is isomorphic to a product $Z \times Y$ where Y is the fiber above one point.

12. THE CURVATURE TENSOR AS A DIFFERENTIAL FORM

Let $\pi: X \to Z$ continue as a fiber space with a connection defined by an $\mathcal{F}(Z)$-linear mapping, a horizontal lifting, of vector fields on Z to vector fields on X. We have defined the curvature form as an $\mathcal{F}(Z)$-bilinear mapping from vector fields on Z to the vertical vector fields on X. Ω has some of the algebraic flavor of a two-differential form. In this section I will discuss how it can be related more precisely to two-differential forms on Z.

There are important physical reasons in back of this. In Maxwell theory the electromagnetic field can be realized as a two-differential form F on a space-time manifold $Z = R^4$, with coordinate labelled (z^u).

$$0 \leq \mu, \nu \leq 3 \; ; \quad 1 \leq i, j \leq 3$$

$$F = F_{\mu\nu} dz^\mu \quad dz^\nu \tag{12.1}$$

$$z^0 = t = \text{time}, \; z^i = \text{space coordinates}$$

$$E = E_{0i} \, dt \wedge dz^i \text{ , the "electric" part} \tag{12.2}$$

$$B = F_{ij} \, dz^i \wedge dz^j \text{ , the "magnetic" part} \tag{12.3}$$

$$F = E + B \, . \tag{12.4}$$

A topic central to the Yang-Mills generalization (and to the Kaluza-Klein theory) is realizing F as a curvature tensor of a connection. Of course, it was also the goal of the older work of Einstein, Weyl, and others in a Unified Gravitation-Electromagnetic theory. I will keep at the general level in this section, and return to the electromagnetic case (and the Yang-Mills generalization) at a later point.

First, let us review material about differential forms. Let M be a manifold.

A *(scalar-valued) two-form* on M is an $\mathcal{F}(M)$-bilinar skew-symmetric map $\omega: \mathcal{V}(M) \times \mathcal{V}(M) \to \mathcal{F}(M)$. Its *value* at $p \in M$ is an R-bilinear, skew-symmetric map:

$$\omega(p): M_p \times M_p \to R \quad .$$

Then, ω can be identified with the tensor field $p \to \omega(p)$. To put this another way, let E be the product bundle $E = M \times R$ over M. Let E(p) (isomorphic to R, of course) be the fiber above the point p. Then, $\omega(p)$ is an R-bilinear skew-symmetric map

$$\omega(p): M_p \times M_p \to E(p) \quad .$$

M_p is the fiber of the tangent bundle. ω then can be defined as a bilinear, skew-symmetric bundle map

$$\omega: T(M) \times T(M) \to E \quad . \tag{12.5}$$

We can now generalize, replacing the product bundle E with an arbitrary vector bundle over M. This leads to the concept of a *differential form on a manifold* M *with values in a vector bundle* E. An alternate algebraic version would be an $\mathcal{F}(M)$-bilinear, skew-symmetric map

$$\omega: \mathcal{V}(M) \times \mathcal{V}(M) \to \Gamma(E)$$

between the modules of *cross-sections* of the vector bundles. Geometrically (and physically) all these concepts are equivalent. Of course, in various applications one formulation is more natural and/or convenient. If one wants to calculate in a *coordinate free* way, the module approach is usually more useful; for calculations in local coordinate systems, the vector bundle approach combined with local product structures for the vector bundles, is usually better.

Let us now return to a general fiber space $\pi: X \to Z$, with connection:

$$h: \mathcal{V}(Z) \to \mathcal{H} \subset \mathcal{G}(\pi)$$

and curvature:

$$\Omega: \mathcal{V}(Z) \times \mathcal{V}(Z) \to VE \quad .$$

We have seen that Ω is $\mathcal{F}(Z)$-bilinear and skew-symmetric. Thus, if there were a vector bundle E over Z such that its space of cross-sections $\Gamma(E)$ was isomorphic (as a $\mathcal{F}(Z)$-module) to the $\mathcal{F}(Z)$-module VE of vertical vector fields, then Ω would be a *two-differential form on* Z *with values in* E.

Now, there certainly is such a bundle, but its fiber is *infinite dimensional*, namely,

$$E = \{(z,W): z \in Z,\ W \in \mathcal{V}(\pi^{-1}(z))\} \quad . \tag{12.6}$$

In words, the fiber above a point $z \in Z$ is the vector space of smooth vector fields on the fiber $\pi^{-1}(z)$ of π above the point z. (Since π is a submersion map, note that $\pi^{-1}(z)$ is a regularly embedded submanifold of X, hence, this vector space is well-defined.) However, to work with this vector bundle in a serious way would involve development of an analytic apparatus, which I do not want to do at the moment. Most of the examples encountered in geometry and physics involve the following special assumptions:

> There is a subbundle E' of E with *finite dimensional fiber* so that Ω takes values in E'. (12.7)

We have seen one example of this: For the flat connection, E' can be taken as the zero bundle. The standard sort of *principal bundle connections* offer another class of examples satisfying (12.7). In this case, E' can be taken as a vector bundle whose fiber is a finite dimensional Lie algebra $\mathcal{K}$, which is the Lie algebra of the structure group K of connections. A special example of this sort is the *affine connection*, i.e., linear connections in the tangent bundle. The curvature takes values in the vector bundle over Z whose fiber is the space of linear maps of the tangent bundle to itself.

We will now review the theory of connections in principal fiber bundles. They play a particularly important role in both mathematical physics and differential geometry.

13. CONNECTIONS IN PRINCIPAL FIBER BUNDLES

Let $\pi: X \to Z$ continue as a fiber space. Let K be a Lie group which acts on X as a transformation group. Suppose this action is *free*, i.e., the following condition is satisfied:

$$\text{If } kx = x, \text{ for } k \in K,\ x \in X, \text{ then } k = \text{identity} \tag{13.1}$$

Suppose also that:

$$K \text{ acts transitively on the fiber of } \pi. \tag{13.2}$$

With these conditions, (X,Z,π,K) is called a *principal fiber bundle* with K *as structure group*.

(13.1) and (13.2) imply that each fiber can be identified with K itself; but not in q unique way. This suggests one way to construct such bundles, as products $Z \times K$:

$$\begin{aligned} X &= Z \times K \\ k(z,k') &= (z,kk') \text{ for } z \in Z;\ k,k' \in K \\ \pi(z,k) &= z \quad . \end{aligned} \tag{13.3}$$

<u>Remark</u>. One can also construct another bundle, by letting K act on the right:

$$\begin{aligned} X' &= Z \times K \\ k(z,k') &= (z,k'k^{-1}) \\ \pi\,(z,k) &= z \quad . \end{aligned} \tag{13.4}$$

These bundles are isomorphic. The isomorphism is

$$(z,k) \to (z,k^{-1}) \tag{13.5}$$

Here is a main geometric property of this type of bundle:

Theorem 13.1. Let $\pi: X \to Z$ be a principal fiber bundle with K as structure group, i.e., let (13.1) and (13.2) be satisfied. The cross-sections $\Gamma(\pi)$ are then in one-one correspondence with the isomorphisms of X with the product bundle (13.3).

Proof. Let $\gamma: Z \to X$ be a cross-section. Define a map

$$\phi_\gamma: Z \times K \to X$$

as:

$$\phi_\gamma(z,k') = k'\gamma(z) \quad . \tag{13.6}$$

Then,

$$\phi_\gamma(k(z,k')) = \phi_\gamma(z,kk')) = kk'\gamma(z) = k\phi_\gamma(z,k') \quad ,$$

i.e., ϕ_γ *intertwines the action of* K.

It is readily seen that ϕ_γ is a diffeomorphism. Q.E.D.

Let (X,Z,π,K) be such a principal bundle with K as structure group. The Lie algebra $\mathcal{K}$ acts on a Lie algebra of vector fields on X. It acts *freely*, i.e., if

$$V(x) = 0, \text{ for } V \in \mathcal{K}, \ x \in X, \text{ then } V = 0 \quad . \tag{13.7}$$

Since $\pi(kx) = \pi(x)$ for $k \in K$, we see that:

$$\pi_*(V) = 0 \text{ for } V \in \mathcal{K} \tag{13.8}$$

i.e., $\quad \mathcal{K} \qquad VE \quad ,$ (13.9)

where VE is the Lie algebra of vectual vector fields. One can then define a linear map

$$\mathcal{F}(X) \otimes \mathcal{K} \to VE$$

as follows:

$$f \times V \to fV \quad . \tag{13.10}$$

Theorem 13.2. The map (13.10) is an isomorphism between the $\mathcal{F}(X)$-module $\mathcal{F}(X) \otimes \mathcal{K}$ and the $\mathcal{F}(X)$-module of vertical vector fields for the principal bundle.

Proof. Let $V_1,\ldots V_m$ be a basis for $\mathcal{K}$. For each $x \in X$, $V_1(x),\ldots,V_m(x)$ is a basis for the vertical vectors. Thus, any vertical vector field V can be written as

$$V = f_1(x)V_1(x) + \ldots + f_m(x)V_m(x) \quad .$$

Since the vector fields $V_1,\ldots,V_m$ on X are linearly independent, the coefficients $x \to f_1(x),\ldots,f_n(x)$ are C^∞ functions, as x ranges over X. Q.E.D.

Of course, the map (13.10) is not a Lie algebra isomorphism between the group algebra $\mathcal{F}(Z) \otimes \mathcal{K}$ and VE. However, consider $\pi^*(\mathcal{F}(X)) \otimes \mathcal{K}$ as a Lie subalgebra of the gauge (Lie) algebra $\mathcal{F}(X) \otimes \mathcal{K}$.

Theorem 13.3. The map (13.10) is a one-one Lie algebra homomorphism from the Lie algebra

$$\pi^*(\mathcal{F}(Z)) \otimes \mathcal{K} \tag{13.11}$$

to the Lie algebra VE. In other words, the Lie algebra VE has a Lie subalgebra isomorphic to the Lie algebra (13.11).

Now, let us consider connections for the principal bundle (X,Z,π,K). Let $H: x \to H(x) \quad X_x$ be a field of horizontal subspaces for π. Then, for each $x \in X$, $X_x = \mathcal{K}(x) \oplus H_x$, where $\mathcal{K}(x)$ denotes the set of values at x of the vector fields in . One can thus define a K-valued one-form θ on X by the following formula:

$$\theta(v) = \text{element of } \mathcal{K} \text{ such that } v = \theta(v) \in H_x \tag{13.12}$$

Thus, $\theta(v) = 0$ if and only if $v \in H_x$. Hence,

$$\theta(h(V)) = 0 \text{ for } V \in \mathcal{V}(Z) \quad .$$

The connection can be defined by the $\mathcal{K}$-valued one-form θ. Of course, we have the alternative of defining the connection as an $\mathcal{F}(Z)$-linear map

$$h: \mathcal{V}(Z) \to \mathcal{G}(\pi)$$

We are now in a position to derive what Cartan called the Structure Equation for the connection. Cartan did this by using bases for the differential forms on X (which he called "moving frames") conveniently adapted to the geometric situation. We shall do this also.

Let us choose indides in the following ranges and the summation convention on these indices:

$$1 \le i,j,k \le m = \dim Z$$

$$m + 1 \le a,b,c \le n = \dim X$$

$$1 \le u,v,w \le n \quad .$$

Let (ω^u) be a basis for one-forms on X such that $\omega^i = 0$ defines the fibers of π. Then,

$$\omega^i(\mathcal{K}) = 0, \text{ and } \omega^a(\mathcal{K}) \text{ is constant} \quad .$$

Theorem 13.4. If $\gamma: Z \to X$ is a cross-section of the fiber space (X,Z,π), i.e., if $\pi\gamma$ = identity, then the one-forms

$$\gamma^*(\omega^1),\ldots,\gamma^*(\omega^m)$$

are a basis for one-forms on the base space Z.

Proof. We use linear algebra. Notice that the number of forms in question is equal to the dimension of Z. Hence, it suffices to prove that they are linearly independent at each point of Z.

Suppose otherwise, i.e., that there is a point $z \in Z$ and nonzero real numbers λ_i such that:

$$\lambda_i\gamma^*(\omega^i)(z) = 0 = \gamma^*(\lambda_i\omega^i(z)) = \lambda_i\omega^i(\gamma_*(Z_z)) \quad . \tag{13.13}$$

γ is a cross-section, hence $1 = \pi_*\gamma_*$. It follows that the tangent space $X_{\gamma(z)}$ is the direct sum of the subspaces $\pi_*^{-1}(0)$ and $\gamma_*(Z_z)$. It follows from (13.13) and the hypothesis that ω^i is zero on the vectors that are tangent to the fibers of π, that

$$\lambda_i\omega^i(X_{\gamma(z)}) = 0 \quad ,$$

which implies that

$$\lambda_i = 0 \quad ,$$

since the one-forms ω^i are, by hypothesis, linearly independent at each point of X. Contradiction. Q.E.D.

Now we can exterior differentiate the forms $(\omega^u) = (\omega^i,\omega^a)$ to obtain what Cartan called the Structure Equations of the Moving Frame:

$$d\omega^u = f^u_{vw}\omega^v \wedge \omega^u \tag{13.14}$$

$$f^u_{vw} + f^u_{wv} = 0 \quad . \tag{13.15}$$

Then (f^u_{vw}) are, of course, functions on X. However, some of them are *constant* (and some of these coordinates are *zero*) as a consequence of the geometry.

First, the Pfaffian system:

$$\omega^i = 0$$

is completely integrable, i.e., defines a foliation on Z. (Its leaves are the fibers of π.) Hence,

$$f^i_{ab} = 0 \quad . \tag{13.16}$$

Second, the ω^a restricted to the fibers of π are (when they are identified with the Lie group K) just the Maurer-Cartan forms of K, i.e.,

$$f^a_{bc} = \text{constant}$$
$$= \text{structure constants of the Lie group K.} \tag{13.17}$$

Let us now rewrite (13.14), taking into account relations (13.16) and (13.17):

$$d\omega^i = f^i_{jk}\omega^j \quad \omega^k + f^i_{ak}\omega^a \quad \omega^k \tag{13.18}$$

$$d\omega^a = f^a_{bc}\omega^b \quad \omega^c + f^a_{ij}\omega^i \quad \omega^j + f^a_{kb}\omega^k \quad \omega^b \tag{13.19}$$

Now, fix a cross-section map $\gamma: Z \to X$. By Theorem 13.4, the $\gamma^*(\omega^i)$ are a basis for one-forms on Z.

Let (V_i) be the dual basis for vector fields on Z,

$$\gamma^*(\omega^i)V_j) = \delta^i_j \quad . \tag{13.20}$$

The $h(V_i)$ are then horizontal vector fields on X, i.e.,

$$\omega^a(h(V_i)) = 0 \quad . \tag{13.21}$$

The curvature tensor Ω is determined by its values on the vertical vectors.

$$\Omega(V_i,V_j) = h([V_i,V_j]) - [h(V_i), h(V_j)]$$

$$\omega(\Omega(V_i,V_j)) = 0$$

Set:

$$\Omega^a_{ij} = \omega^a(\Omega(V_i,V_j)) \quad . \tag{13.22}$$

They are the *components* of the curvature tensor in this moving frame. Notice that they are functions on X. They do depend on the choice of cross-section γ. (γ is the "moving frame.")

<u>Theorem 13.5</u>. The vector fields $h(V_i) = V_i'$ on X are the dual to ω^i, i.e.,

$$\omega^a(V_i') = 0 \tag{13.23}$$

$$\omega^j(V_i') = \delta^j_i \quad . \tag{13.24}$$

<u>Proof</u>. These are just relations (13.20) and (13.21) rewritten.

For a point x of X, to calculate Ω^a_{ij}, we can take advantage of the freedom of choice in the moving frame. Namely, we can, without loss in generality, suppose that the (ω^i) satisfy:

$$d\omega^i = 0 \quad \text{at } x \quad . \tag{13.25}$$

(What this means is that, if (ω^i) do not satisfy (13.25), a new set (ω^i_1) can be chosen so that

$$d\omega^i_1 = 0 \text{ at } x \tag{13.26}$$

$$\omega^i_1(x) = \omega^i(x) \quad .$$

This trick (which is a version of the method of "normal coordinates" and "moving frame") often dramatically simplies computation.)

Let $z = \pi(x)$. (13.26) implies that

$$[V_i,V_j](x) = 0 \quad . \tag{13.27}$$

Hence,

$$\Omega(V_i,V_j)(x) = -[h(V_i),h(V_j)](x) = -[V'_i,V'_j](x) \quad .$$

Hence,

$$\Omega^a_{ij} = -\omega^a([V'_i,V'_j]) = \text{, using (13.23),} \quad d\omega^a(V'_i,V'_j)$$

$$= \text{, using (13.19),} \quad 2f^a_{ij}$$

We have now proved:

<u>Theorem 13.6.</u> The structure relations determine the curvature tensors. Explicitly, we have:

$$d\omega^i = f^i_{jk}\omega^j \wedge \omega^k + f^i_{ak}\omega^a \wedge \omega^k \tag{13.28}$$

$$d\omega^a = f^a_{bc}\omega^b \wedge \omega^c + \frac{1}{2}\Omega^a_{ij}\omega^i \wedge \omega^j + f^a_{kb}\omega^k \wedge \omega^b \tag{13.29}$$

Let us examine the third term on the right hand side of (13.29). Its vanishing expresses an important property of the connection.

Note that the Pfaffian equations $\omega^i = 0$ define the horizontal vectors, which, of course, determine the connection. Let V be a vector field in $\mathcal{K}$. It satisfies the following conditions:

$$\omega^i(V) = 0 \tag{13.30}$$

$$\omega^a(V) = \text{constant} \tag{13.31}$$

Also, we have:

$$\mathcal{L}_V(\omega^a) = d(V\,\omega^a) + V\,d\omega^a \quad . \tag{13.32}$$

But, the first term on the right hand side of (13.32) vanishes, because of (13.31). Using also (13.29), we have:

$$\mathcal{L}_V(\omega^a) \equiv -f^a_{kb}\omega^b(V)\omega^k \text{ mode } (\omega^a) \quad . \tag{13.33}$$

We have proved:

<u>Theorem 13.7.</u> The horizontal Pfaffian system H is invariant under the action of the structure group K if and only if the following relations are satisfied:

$$f^a_{kb} = 0 \quad . \tag{13.34}$$

Thus, in trying to simplify the structure relations for the connection (following Cartan's technique) we encounter in a natural way the conditions Ehresmann imposed [17], namely the invariance of hte horizontal Pfaffian system under the structure group of the bundle. Let us call these K-*invariant connections*.

Recapitulation of the Basic Definition. Let $\pi: X \to Z$ be a principal K-bundle, with a Lie group K acting freely on X and transitively on the fibers of π. A connection for the fiber space, defined by a field $x \to H(x)$ of horizontal tangent subspaces, is said to be K-*invariant* if the following condition is satisfied:

$$k_*(H_x) = H_{kx} \quad \text{for all } k \in K,\ x \in X \ . \tag{13.35}$$

We can now sum up the "moving frame" relations in the following, more precise way, keeping account of the global situation.

Theorem 13.8. Let (X,Z,π,K) be a principal K-bundle, with a K-invariant connection (K a connected Lie group, defined by a field $x \to H(x)$ of horizontal tangent subspaces.) Let $\mathcal{K}$ be the Lie algebra of K, realized as a Lie algebra of vector fields on X. (ω^i,ω^a) be a basis of one-forms on an open subset U of X satisfying the following relations:

$$\omega^a(\mathcal{K}) = \text{constant} \tag{13.36}$$

$$\omega^a(H_x) = 0 \tag{13.37}$$

$$\mathcal{K} \lrcorner\, \omega^i = 0 \tag{13.38}$$

Then, there are differential forms in U (ω^i_j,Ω^a) such that:

$$d\omega^i = \omega^i_j \wedge \omega^j \ , \tag{13.39}$$

$$d\omega^a = \Omega^a + \lambda^a_{bc}\omega^b \wedge \omega^c \tag{13.40}$$

with the following conditions also satisfied:

$$\mathcal{K} \lrcorner\, \Omega^a = 0 \tag{13.41}$$

λ^a_{bc} are the structure constants of the Lie algebra $\mathcal{K}$, i.e., the f^a_{bc} are constant on the fibers of π. (13.42)

Proof. Start from relations (13.28)-(13.29). Set

$$\omega^i_j = - f^i_{jk}\omega^k + f^i_{aj}\omega^a \tag{13.43}$$

$$\Omega^a = \frac{1}{2}\Omega^a_{ij}\omega^i \quad \omega^j \ . \tag{13.44}$$

Note that (13.41) is satisfied using relation (13.38). To prove (13.42), let (V_a) be a basis for $\mathcal{K}$. Then,

$$[V_a,V_b] = \lambda^c_{ab}\, V_c \ ,$$

where λ^c_{ab} are the *structure constants* of the Lie algebra $\mathcal{K}$. Hence,

$$d\omega^a(V_b,V_c) = V_b(\omega^a(V_c)) - V_c(\omega^a(V_b)) - \omega^a([V_b,V_c])$$

$$= 0 - 0 - \lambda^a_{bc}$$

Q.E.D.

Of course, we see again from (13.40) why the vanishing of the curvature tensor, which is essentially Ω^a, determines the "flatness" of the connection, i.e., the complete integrability of the field $x \to H_x$ of horizontal tangent subspace (or, dually, the Pfaffian equations $\omega^a = 0$), which determine the connection.

14. LIE ALGEBRA-VALUED DIFFERENTIAL FORMS AND THE MAURER-CARTAN EQUATIONS

We have seen that Equations (13.39-40) are the basic structure equations of a K-invariant connection on a principal fiber bundle with $\mathcal{K}$ as structure group. I now want to pause in the development of connection-gauge field-Yang-Mills theory to review some relatively elementary and standard material in differential geometry and Lie group theory associated with the "Maurer-Cartan" equations.

Let X be a manifold. We have already defined vector-bundle valued differential forms on X. In particular, vector-space valued forms are a special case, where the vector bundle E is the product $E = X \times V$ of X with a vector space V. Of course, such a form can be defined directly as a multilinear mapping

$$\theta: T(X) \times \ldots \times T(X) \to V \quad .$$

In connection theory (and a fortiori in a gauge-field-Yang-Mills theory) we are interested in the special case that

$$V = \mathcal{K},$$

the Lie algebra of a connected Lie group K.

We have to be more specific about the definition of $\mathcal{K}$. The most natural choice is to say that it is *defined* as the set of one-parameter subgroups of K. Thus, for $V \in \mathcal{K}$

$$t \to \exp(tV)$$

is by *definition*, the one-parameter subgroup of K associated with it.

K acts freely in two ways on itself, by *left* and *right* translation:

$$L_k(k') = kk'$$

$$R_k(k') = k'k^{-1} \quad .$$

The following result (which we will not prove here) is, in a sense, the main theorem of Lie group theory.

<u>Theorem 14.1</u>. $\mathcal{K}$ can be made into a Lie algebra

$$(V_1, V_2) \to [V_1, V_2]$$

with the following properties:

a) There are Lie algebra isomorphisms

$$\alpha_R: \mathcal{K} \to \mathcal{V}(\mathcal{K})$$

$$\alpha_L: \mathcal{K} \to \mathcal{V}(\mathcal{K})$$

b) For $v \in \mathcal{K}$, $\alpha_L(v)$ (resp. $\alpha_R(v)$) is the infinitesimal generator of the one-parameter group of diffeomorphisms

$$t \to L_{\exp}(tV) \qquad (\text{resp. } t \to R_{\exp}(tV))$$

c) $[\alpha_R(\mathcal{K}), \alpha_L(\mathcal{K})] = 0$

d) If $V \in \mathcal{V}(K)$ satisfies:

$$[V, \alpha_R(\mathcal{K})] = 0 \quad ,$$

then

$$V \in \alpha_L(\mathcal{K}) \quad .$$

e) Relation (d) holds if L and R are permuted.

f) For each $k \in K$, the maps

$$v \to \alpha_R(v)(k)$$

$$v \to \alpha_L(v)(k)$$

are vector space isomorphisms of $\mathcal{K}$ with the tangent space K_k to K at k.

g) If (V_a), $1 \leq a \leq n$, is a basis of $\mathcal{K}$, then

$$\alpha_R(V_a) \text{ and } \alpha_L(V_a)$$

are both $\mathcal{F}(K)$-module bases of $\mathcal{V}(K)$. (Such bases are called, in differential geometry, *absolute parallelisms*.) Thus, if

$$[V_a, V_b] = \lambda^c_{ab} V_c \quad ,$$

with *constant* λ (called the *structure constants* of the Lie algebra for the given bases), then

$$[\alpha_L(V_a), \alpha_L(V_b)] = \lambda^c_{ab}\alpha_L(V_c)$$

$$[\alpha_R(V_a), \alpha_R(V_b)] = \lambda^c_{ab}\alpha_R(V^c)$$

h) The basis (ω^a_L) of one-forms on K dual to the $\alpha_L(V_a)$, i.e., the one-forms which satisfy the following relation:

$$\omega^a_L(\alpha_L(V_b)) = \delta^a_b \quad ,$$

Also satisfy the following relations:

$$d\omega^a_L = \frac{1}{2}\lambda^a_{bc}\,\omega^b_L \quad \omega^c_L \tag{14.1}$$

Of course, also

$$d\omega^a_R = \frac{1}{2}\lambda^a_{bc}\,\omega^b_R \quad \omega^c_R$$

with

$$\omega^a_R(\alpha_R(V_b)) = \delta^a_b \quad .$$

i) The (ω^a_L) are invariant under *left translation*:

$$L^*_k(\omega^a_L) = \omega^a_L \quad \text{for } k \in K$$

$$\mathcal{L}_{\alpha_R(V)}\,(\omega^a_L) = 0 \text{ for } v \in \mathcal{K} \quad .$$

Equations (14.1) considered as a set of differential equations for the one-forms ω^a_L, are called the *Maurer-Cartan equations*. They were Cartan's favorite tool in the development of differential geometry and Lie group theory. (Lie preferred vector fields.)

We can now define $\mathcal{K}$-valued one-forms on K, called the *Maurer-Cartan forms*

$$\omega_L = \omega_L^a V_a$$

$$\omega_R = \omega_R^a V_a$$

These objects are easy to calculate if K is exhibited as a group of matrices, say:

$$K \subset GL(N,R) \quad .$$

Then,

$$\omega_L = k^{-1} dk$$

$$\omega_R = dk\, k^{-1} \quad . \tag{14.2}$$

Now, there is a notational convention inherent in these formulas. It is necessary to identify $\mathcal{K}$ with a set of matrices, and to identify the tangent bundle

$$(T(K)$$

with

$$K \times \mathcal{K} \quad ,$$

i.e., a pair (k,M) of a $k \in K$ and an $N \times N$ matrix M. These identifications are confusing, and it is doing these computations (which, of course, are necessary in correlating the coordinate-free geometer's notation with the physicist's) that the beginner most often gets confused. Here are some examples.

Example 1. $K = O(N,R)$, the $N \times N$ real orthogonal matrices. K is then the set of $M \in GL(N,R)$ such that

$$MM^T = 1 \quad .$$

(M^T is the transpose of M. 1 is the identity matrix.) Then,

$$0 = dMM^T + MdM^T \quad . \tag{14.3}$$

Since $M^T = M^{-1}$, (14.2) is:

$$dMM^{-1} + (dMM^{-1})^T = 0 \quad . \tag{14.4}$$

Thus, if we *define* $\mathcal{K}$ as the Lie algebra (under matrix commutation) of $N \times N$ skew-symmetric matrices, then

$$\omega_R = dMM^{-1} \tag{14.5}$$

is the *right*-invariant Cartan-Maurer form. Relation (14.4) says that ω_R is a one-form on K with values in $\mathcal{K}$, the skew-symmetric matrices. The other "classical" matrix groups can be handled by this technique.

We can obtain the equations satisfied by ω_L and ω_R by applying exterior derivatives to both sides of (14.2):

$$d\omega_L = d(k^{-1}) \quad dk \quad .$$

Now,

$$0 = d(kk^{-1})$$
$$= (dk)k^{-1} + kd(k^{-1}) \quad ,$$

or

$$d(k^{-1}) = -k^{-1}dk\, k^{-1}$$

$$d\omega_L = k^{-1}dk \wedge k^{-1}dk = -\omega_L \wedge \omega_L \quad . \tag{14.6}$$

Now, in (14.6), the symbol $\wedge$ denotes exterior product of *matrices* of differential forms. For example, in the 2×2 case

$$\omega = \begin{pmatrix} \omega_{11} & \omega_{12} \\ \omega_{21} & \omega_{22} \end{pmatrix}$$

$$\omega \wedge \omega = \begin{pmatrix} \omega_{11} & \omega_{12} \\ \omega_{21} & \omega_{22} \end{pmatrix} \wedge \begin{pmatrix} \omega_{11} & \omega_{12} \\ \omega_{21} & \omega_{22} \end{pmatrix}$$

$$= \begin{pmatrix} \omega_{12} \wedge \omega_{21} & \omega_{12} \wedge \omega_{12} + \omega_{12} \wedge \omega_{22} \\ \omega_{21} \wedge \omega_{11} + \omega_{22}\,\omega_{21} \, , & \omega_{21} \wedge \omega_{12} \end{pmatrix}$$

Let us now interpret (14.6) in terms of abstract Lie algebra theory. ω_L is a $\mathcal{K}$-valued one-form on K. For $V_1, V_2 \in \mathcal{V}(K)$

$$d\omega_L(V_1,V_2) = V_1(\omega_L(V_2)) - V_2(\omega_L(V_1)) - \omega_L([V_1,V_2])$$

$$(\omega_L \quad \omega_L)(V_1,V_2) = \omega_L(V_1)\omega_L(V_2) - \omega_L(V_2)\omega_L(V_1)$$

$$= [\omega_L(V_1), \omega_L(V_2)] \, .$$

Hence, (14.6) is equivalent to the following relation:

$$D\omega_L = 0 \tag{14.7}$$

where:

$$D\omega_L(V_1,V_2) = V_1(\omega(V_2)) - V_2(\omega(V_1)) - \omega_L([V_1,V_2]) - [\omega_L(V_1), \omega_L(V_2)] \tag{14.8}$$

We can now abstract some useful definitions for Lie algebra-valued differential forms:

<u>Definition</u>. Let $\mathcal{K}$ be a Lie algebra, X a manifold. Let

$$\omega: T(X) \to \mathcal{K}$$

be a $\mathcal{K}$-valued one-form on X. Then, $D\omega$ is a $\mathcal{K}$-valued two form on X, defined by the following formula:

$$D\omega(V_1,V_2) = V_1(\omega(V_2)) - V_2(\omega(V_1)) - \omega([V_1,V_2]) - [\omega(V_1), \omega(V_2)] \tag{14.9}$$

We now write the Maurer-Cartan equations in terms of this operation.

<u>Theorem 14.2</u>. Let K be a Lie group and let (V^a) be a basis of $\mathcal{K}$. Let (ω_a) be a dual basis of the left (or right) invariant one-forms on K.

Let

$$\omega = \omega_a V^a$$

be *the* $\mathcal{K}$-valued left invariant one-form on K. Then

$$D\omega = 0 \, , \tag{14.10}$$

where D is the operation give by formula (14.9).

<u>Proof</u>. Since $D\omega$, as given by (14.9), is obviously left-invariant, it suffices to prove (14.10) at one point. To do this, it suffices to prove that

$$D\omega(V^a,V^B) = -\omega([V^a,V^b]) - [\omega(V^a),\omega(V^b)]$$
$$= \omega_c([V^a,V^b])V^c - [\omega_c(V^a)V^c, \omega_c(V^b)V^c]$$
$$= [V^a,V^b] - [V^a,V^b] = \underline{\underline{0}} ,$$

which proves (14.10), hences also Theorem 14.2.

Theorem 14.3. Let X be a manifold and let ϕ: X → K be a map. Let

$$\theta: T(X) \to \mathcal{G}$$

be the one-form which is pull-back of the $\mathcal{K}$-valued Maurer-Cartan form ω on K, i.e., $\theta = \phi^*(\omega)$, where ω is the Maurer-Cartan form D + K. Then,

$$D\theta = 0 \tag{14.10}$$

where D is the operator given by (14.9).

Proof. Obviously, the D operation is "covariant," i.e., preserved under the pull-back map ϕ^*.

Now, for the converse.

Theorem 14.4. Suppose X is a manifold with $\mathcal{K}$-valued one-form θ on X such that:

$$D\theta = 0 . \tag{14.11}$$

Then, each point $x \in X$ has an open neighborhood U and a map

$$\phi_U: U \to K$$

such that:

$$\phi_U^*(\omega) = \theta , \tag{14.12}$$

where ω is the (left-invariant $\mathcal{K}$-valued) Maurer-Cartan form on K.

Proof. Construct:

$$Y = X \times K. \tag{14.13}$$

Consider the Pfaffian system

$$\theta - \omega$$

on Y. As a consequence of (14.10), it is completely integrable (see Remark below). Consider a leaf L of this folation. It is readily seen that the projection

$$L \to X$$

is a local diffeomorphism. Choosing a local inverse

$$X \quad U \to L \quad X \times K ,$$

following by the projection

$$X \times K \to K$$

provides the map ϕ_U.

Remark. I will make this argument more explicit in case K is a matrix group, since it is quite important in the theory of nonlinear waves. Suppose

$$X = R^n .$$

Suppose

$$K \subset GL(N,R) ,$$

i.e., an element of K (or $\mathcal{K}$) is an N × N real matrix. Let

$$(x^i) \quad , \quad 1 \leq i,j \leq n$$

be Cartesian coordinates on R^n. Then

$$\theta = A_i dx^i \quad ,$$

where $A_1(x),\ldots,A_n(x)$ are $N \times N$ matrices, which are elements in Condition (14.11 is:

$$\partial_i(A_j) - \partial_j(A_i) = [A_i,A_j] \qquad (14.14)$$

We are looking for k ($\in \mathcal{K}$) as a function of x, so that

$$\theta = k^{-1}dk \quad , \qquad (14.15)$$

i.e.,

$$\partial_i(k) = kA_i \quad . \qquad (14.16)$$

We see that (14.14) are the *integrability conditions* for the differential equations (14.16). (k as a function of x can be determined by solving a set of ordinary, linear equations depending on parameters.)

For example, set

$$k(t) = k(at) \quad , \quad a \in R^n \quad .$$

Then,

$$\frac{dk}{dt} = a^i A_i(at)k(t) \quad . \qquad (14.17)$$

These linear equations can be solved for all t. One sees (using a classical argument, that in modern terms is essentially the Frobenius theorem) that the solutions of (14.16) exist over *all* of R^n. We will now generalize it.

Return to the case of a general manifold X and a connected Lie group K, and a $\mathcal{K}$-valued one-form on X, θ such that

$$D\theta \equiv d\theta + [\theta,\theta] = 0 \quad . \qquad (14.18)$$

Then, X can be covered by open sets

$$\{U\} \quad ,$$

that, for each $U \in \{U\}$, there is a map $\phi_U\colon U \to K$ such that:

$$\phi_U^*(\omega) = \theta \quad , \qquad (14.19)$$

where ω is the $\mathcal{K}$-valued Maurer-Cartan form on K.

One can arrange things (general theorem about paracompact connected manifolds) so that the covering $\{U\}$ has the following property:

Whenever $U,U' \in \{U\}$ intersect, the intersection $U \cap U'$ is *connected*. (14.20)

Then, one sees (since the solution ϕ_U of (14.19) is unique up to left translation) that, on each $U \cap U'$,

$$\phi_U = k_{UU'}\phi_{U'} \quad , \qquad (14.21)$$

where $(k_{UU'})$ is an element of K. Whenever $U,U',U'' \in \{U\}$ intersect

$$k_{UU''} = k_{UU'}k_{U'U} \quad . \qquad (14.22)$$

Condition (14.21) is a "Cech cocycle condition." If X is simply connected, a standard argument gives the existence of a set $\{k_U\}$ of elements of K, such that, whenever U U' is non-empty,

$$k_U = k_{UU'}k_{U'}\quad .$$

Thus,

$$(k_U^{-1}\phi_U) = \phi$$

is globally defined on X', and gives a solution to the Maurer-Cartan equations

$$\phi^*(\omega) = \theta\quad .$$

If X is not simply connected, there is a subgroup of the fundamental group $\pi_1(X)$ of X which is the "obstruction" to the *global* existence of ϕ.

15. RIEMANNIAN GEOMETRY, AND GENERALIZED KLEIN GEOMETRIES, BY CARTAN'S METHODS

The main purpose of this survey paper is to show how the fundamental work of Cartan and Ehresmann is involved in the physics of Yang-Mills fields. Of course physics was not the main motivation for Cartan. (Although, after Einstein's development of General Relativity Cartan did devote considerable effort to exploring the geometric foundations of physics!) His goal was to develop ideas of "generalized geometries" based on a group, following Felix Klein's "Erlanger Program." Thus, he developed Euclidean, affine, projective, conformal,... connections, corresponding to the classical geometries.

This point of view is so important for understanding Cartan's methodology that I want to insert a relatively brief discussion of it at this point, concentrating on the Euclidean case.

Let V be a real vector space of dimension n, with a positive-definite symmetric bilinear form

$$(v_1,v_2) \to \langle v_1,v_2\rangle$$

defined on V. Choose indices, with the summation convention as follows:

$$1 \le i,j,k \le n\quad .$$

Euclidean frame for V (with the basic Euclidean geometry defined by the form $\langle\,,\rangle$) is an (n+ 1) tuple:

$$e = (v_0,v_1,\dots,v_n) \tag{15.1}$$

of elements of V such that:

$$\langle v_i,v_j\rangle = \delta_{ij}$$

Let E be the set of such frames and let π: E → V be the map defined by the following formula:

$$\pi(v_0,v_1,\dots,v_n) = v_0\ . \tag{15.2}$$

Denote an element e ∈ E of form (15.1) "vectorially"

$$e = (v_0,\underline{v})$$

$$\underline{v} = (v_1,\dots,v_n)$$

Let K be the Lie group O(n,R) of N × N orthogonal matrices:

$$k = (k_{ij}) \quad , \quad k_{ij}k_{jk} = \delta_{ik}$$

Let K act on E as follows:

$$ke = (v_0, \underline{v}k^{-1}) \ . \tag{15.3}$$

The following result is readily proved, either directly or using general theory, and is left to the reader.

Theorem 15.1. The map π: E → V and the action (15.3) of K define E as a principal K-bundle, with V as base space.

Now, define the following V-valued functions, i.e., zero-degree V-valued differential forms, P, $e_1,\dots,e_n$, as follows:

$$\begin{aligned} P(v_0;v_1,\dots,v_n) &= v_0 \\ e_1(v_0;v_1,\dots,v_n)z &= v_1 \\ &\vdots \\ e_n(v_0;v_1,\dots,v_n) &= v_n \end{aligned} \tag{15.4}$$

Notice that:

$$\langle e_i, e_j \rangle = \delta_{ij} \tag{15.5}$$

I will now state, without proof, a sequence of theorems, whose proof is straightforward, hence is left to the reader.

Theorem 15.2. There exist V-valued one forms (ω_i, ω_{ij}) on E such that the following conditions are satisfied:

$$dP = \omega_i e_i \tag{15.6}$$

$$de_i = \omega_{ij} e_j \tag{15.7}$$

These one-forms are a basis for the F(E)-module $\mathcal{D}^1$(E), i.e., define an absolute parallelism for E.

Theorem 15.3. The Pfaffian system

$$\omega_i = 0 \tag{15.8}$$

on E is completely integrable; its leaves are the fiber map π. Here, the Pfaffian system

$$\omega_{ij} = 0 \tag{15.9}$$

defines the horizontal vectors for an Ehresmann connection for the map π.

In order to prove the integrability of the Ehresmann connection defined by the Pfaffian system (15.9), let us exterior differentiate both sides of (15.6) and (15.7).

$$\begin{aligned} 0 &= d\omega_j e_i - \omega_j \wedge de_j \\ &= d\omega_j e_i - \omega_j \wedge \omega_{ij} e_j \\ &= (d\omega_j - \omega_j \wedge \omega_{ij}) e_j \quad , \end{aligned}$$

which implies:

$$d\omega_j = \omega_{ij} \wedge \omega_j \tag{15.10}$$

Now, exterior differentiate both sides of (15.7)

$$0 = d\omega_{ij} e_j - \omega_{ij} \wedge de_j$$

$$= d\omega_{ij} e_j - \omega_{ij} \wedge \omega_{jk} e_k$$

$$= (d\omega_{ik} - \omega_{ij} \wedge \omega_{jk}) e_k \, ,$$

which implies:

$$d\omega_{jk} = \omega_{ij} \wedge \omega_{jk} \tag{15.11}$$

There is a final relation obtained by exterior differentiating (15.5):

$$0 = \langle de_{ji} e_j \rangle + \langle e_{ij} de_j \rangle$$

$$= \langle \omega_{jk} e_k, e_j \rangle + \langle e_i, \omega_{jk} e_k \rangle$$

$$= \omega_{jk} \delta_{kj} + \omega_{jk} \delta_{ik} \, ,$$

or

$$\omega_{ij} + \omega_{ji} = 0 \tag{15.12}$$

Equation (15.10)-(15.12) are (in Cartan's language) the *Structure Equations of Euclidean Geometry*. (15.12) says that (ω_{ij}) is a one-form on E with values in the Lie algebra of K, = 0(n,R, the skew-symmetric, N × N real matrices. (15.10) expresses the complete integrability of the Pfaffian system (15.8), which we already know since the fibers of π are the leaves. (15.11) is the relation which has the most non-trivial information. It expresses the fact that the curvature of the connection determined by the Pfaffian system (15.9) is *zero*, and that the collection

$$(\omega_i, \omega_{ij})$$

of one-forms defines a $\mathcal{G}$-valued one form on E, with

$$G = \text{group of rigid motion of } R^n \, ,$$

which satisfies the *Maurer Cartan equation*. Then, there are isomorphism, i.e.

$$\begin{array}{ccc} E & \leftrightarrow & G \\ \downarrow & & \downarrow \\ R^n & \leftrightarrow & G/K \end{array}$$

which carry the Maurer-Cartan forms on $\mathcal{G}$ back to the (ω_i, ω_{ij}). Of course the isomorphism in this case is easy to construct using any orthonormal basis of V.

So far Cartan's point of view has merely rephrased standard Euclidean geometry, as it is expressed in Klein's "Erlanger Program" as the theory of the coset space G/K. However, the brilliance of Cartan enters in his concept of a "generalized Euclidean Geometry obtained by modifying the Structure Equations (15.10-15.11) in a way that we shall now describe.

Let $\pi: E \to X$ be a submersion map with:

$$\dim X = n; \ \dim E = \frac{n(n+1)}{2}.$$

Suppose

$$(\omega_i, \omega_{ij}); \ \omega_{ij} + \omega_{ji} = 0$$

are a collection of one-forms on E which form an $\mathcal{F}(E)$-basis of $\mathcal{D}^1(E)$, i.e., which are linearly independent at each point of E. Suppose that the following conditions are satisfied:

$$d\omega_i = \omega_{ij} \wedge \omega_j + T_{ij}\omega_j \tag{15.13}$$

$$d\omega_{ij} = \omega_{jk} \wedge \omega_{kj} + R_{ijke}\omega_k \wedge \omega_e \tag{15.14}$$

where (T_{ij}), (R_{ijke}) are functions on E, which are, for the moment unspecified.

Notice that (15.13) implies that the Pfaffian system $\omega_i = 0$ is completely integrable. Let us suppose that the following conditions satisfied:

The leaves of the foliation $\omega_i = 0$ are the fibers of the map .

<u>Theorem 15.4</u>. There is one, and only one, Riemannian metric β on X such that:

$$\pi^*(\beta) = \omega_j \cdot \omega_j \quad , \tag{15.15}$$

whereas $\cdot$ denotes symmetric product of differential forms.

<u>Proof</u>. Suppose for the moment that a Riemannian metric (i.e., a symmetric positive definite inner product on tangent vectors) β exists, satisfying (15.15). $\pi^*(\beta)$ is, in general, the symmetric bilinear form on tangent vectors to E such that:

$$\pi^*(\beta)(v_1, v_2) = \beta(\pi_*(v_1), \pi_*(v_2)) \tag{15.16}$$

for v_1, $v_2 \in T(E)$.

Let

F $e \in F$,

let

V(e) = set of vertical tangent vectors to E

$= \{v \in E_e : \pi_*(v) = 0\}$

$= \{v \in E_e : \omega_i(v) = 0\}$

H(e) = set of horizontal tangent vectors to E

$= \{v \in E_e : \omega_{ij}(v) = 0\}$

Then,

$$E_e = V(e) + H(e) \quad . \tag{15.17}$$

Thus, $\pi_*(\beta)$ is determined by the values on E(R) where it agrees with $\omega_i \omega_i$. Since this holds for all $e \in E$, and the map $\beta \to \pi_*(\beta)$ is one-one, if β exists, it is unique. Existence of β can also be readily proved..

Let FR(X,β) be the bundle of orthonormal tangent frames for the Riemannian metric on X determined by β. A point of FR(X,β) is then an (n+1)-tuple $(x; v_1, \ldots, v_n)$.

$x \in X$:

$v_1, \ldots, v_n \in X_x$

$\beta(v_i, v_j) = \delta_{ij}$

Now, define a map

$u: E \to FR(X)$

as follows:

For $e \in E$, construct the basis

(u_i, u_{ij}), $u_{ij} + u_{ji} = 0$

to the tangent space of E at e dual to the

(ω_i, ω_{ij}):

$\omega_{ij}(u_k) = 0 = \omega_i(u_{jk})$

$\omega_i(u_k) = \delta_{ik}$

$\omega_{ij}(u_{ke}) = \delta_{ik}\delta_{je}$

Let:

$$u(e) = (\pi(x), \pi_*(u_i)) \tag{15.18}$$

Theorem 15.5. u defined by (15.18) is an isomorphism between E and the bundle FR(X). There is a Cartan connection for FR(X) or the sense of Ehresmann [17,74] and u^k cover both these connection forms into the (ω_i, ω_{ij}) on E. In other words, the "Structure Equations" (15.13)-(15.14) can serve to define a "Cartan connection," as Ehresmann defined it in 1950. The T and R go under u^{-1} into the torsion and curvature of the Cartan connection on FR(X,β), respectively.

The details of the proof are rather lengthy, hence will be deferred to another publication.

The main point I want to make here is that Cartan's approach to a "Generalized Euclidean Geometry" is equivalent to Ehresmann's, but is geometrically and physically more motivated. It can also be generalized to other "geometries" based on other homogeneous spaces G/K, as we now indicate.

Let G be a connected Lie group of dimension m, K a closed subgroup of G of dimension m-n. Let $\mathcal{G}$ be the Lie algebra of G, $\mathcal{K}$ the subalgebra corresponding to the subgroup K.

Let (θ^a, θ^i), $1 \le i,j \le n$; $n+1 \le a, b \le m$,

be a basis of the dual vector space of $\mathcal{G}$, such that:

$$\theta^i(\mathcal{K}) = 0 \tag{15.19}$$

As we have seen, the

(θ^i, θ^a)

can be interpreted as Maurer-Cartan forms in G, where they satisfy structural relation of the following type, with constant coefficient λ:

$$d\theta^j = (\lambda^i_{jk}\theta^j + \lambda^i_{ak}\theta^a) \wedge \theta^k \tag{15.20}$$

$$d\theta^a = \lambda^a_{ij}\theta^i \wedge \theta^j + \lambda^i_{aj}\theta^a \wedge \theta^j + \lambda^i_{bc}\theta^b \wedge \theta^c \tag{15.21}$$

Now, Cartan's notion of a *generalized geometry* associated with the Klein geometry on G/K (in the same sense that Riemannian geometry is the "generalized Euclidean geometry" in Klein's sense) can be made precise by giving the following data:

a) Manifolds E and X with a submersion map $\pi: E \to X$

b) dim E = dim G
dim X = dim (G/K)

c) An F(E)-module basis of $\mathcal{D}^1$ (E) labelled:
(ω^i, ω^a)

d) The Pfaffian system $\omega^i = 0$ is completely integrable. The leaves are the fibers of π.

e) The following Structural Equations are satisfied:

$$d\omega^i - (\lambda^i_{kj}\omega^k + \lambda^i_{aj}\omega^a) \wedge \omega^j = T^i_{jk}\omega^j \wedge \omega^k \qquad (15.22)$$

$$d\omega^a - (\lambda^a_{kj}\omega^k + \lambda^a_{bj}\omega^b) \wedge \omega^j = R^a_{jk}\omega^j \wedge \omega^k \qquad (15.23)$$

$(T^i_{jk}($ is the *torsion*, (R^a_{jk}) is the *curvature*.

Cartan would think of these Structure Equations holdings *locally* in open subsets of X, pulled back via cross-section maps

$\gamma: U \to E$

defined over open subsets U of X. This is why differential forms,rather than vector fields, are preferable as the analytical apparatus of Cartan's ideas - that they "pull back" under maps since they are "covariant" tensor fields enables him to get away with being imprecise about which spaces they are actually defined on.

There are two main classical examples for which Cartan's ideas work brilliantly, the *projective* and *conformal* connections. In the former case,

G = SL(n+1),K), K is the subgroups such that

$$G/K = P_n(R), \qquad (15.24)$$

the n dimensional projective space of one-dimensional linear subsapces of R^{n+1}.

For the conformal case:

G = SO(n+1, 1), the group of (n+2) × (n+2) real matrices whih leave invariant a quadratic form on R^{n+2} whose normsl form has (n+1)-plus signs and 1 minus sign. K is the subgroup such that $G/K = S^n$, the n-sphere in R^{n+1} and such that G acts on S^k as group of *conformal* transformations of the constant-curvative Riemannian nature on S^n.

Here are the Structure Equations, for these two cases, in the form presented in "Interdisciplinary Mathematics," vol. 10.

Projective Connection. The connection is determined by a basis of forms on E (which is of dimension $(n+1)^2 - 1 = n(n+2) = n^2 + 2n$) labelled $\omega^i; \omega^i_j; \omega^0_i$ $1 \leq i,j \leq n$. $\omega^i = 0$ is the foliation determining the fibers of the submersion map $\pi: E \to X$.

$$\begin{aligned} d\omega^i &= \omega^i_j \wedge \omega^j + T^i_{jk}\omega^j \wedge \omega^k \\ d\omega^i_j &= \omega^i_k \wedge \omega^k_j - \omega^0_j \wedge \omega^i - \omega^0_k \wedge \omega^k \delta^i_j + R^i_{jke}\omega^k \wedge \omega^0 \\ d\omega^0_i &= \omega^0_j \wedge \omega^j_i + S_{ijk}\omega^j \wedge \omega^k \end{aligned} \tag{15.25}$$

Conformal Connection. E is now of dimension

$$\frac{(n+2)(n+1)}{2} = \frac{n^2}{2} + \frac{3n}{2} + 1 \tag{15.26}$$

The basis of $\mathcal{D}^1(E)$ is determined by forms labelled:

$$\omega^i; \omega^j_i; \omega_i, \omega$$

$$\omega^j_i + \omega^i_j = 0 \quad , \quad 1 \leq i,j, \leq n \quad .$$

There are

$$1 + 2n + \frac{n(n-1)}{2} = 2n - \frac{n}{2} + \frac{n^2}{2} + 1$$

such independent forms, which of course agree with (15.26).

The Structure Equations for Conformal Connections are now (following Cartan [6], p. 179)

$$\begin{aligned} d\omega^i &= \omega^i_j \wedge \omega^j + T^i_{jk}\omega^j \wedge \omega^k \\ d\omega^i_j &= \omega^i_k \wedge \omega^k_j + R^i_{jke}\omega^k \wedge \omega^e \\ d\omega_i &= \omega \wedge \omega_i + \omega_j \wedge \omega^j_i + S_{ijk}\omega^j \wedge \omega^k \\ d\omega &= \omega_i \wedge \omega^i + T_{ij}\omega^j \wedge \omega^k \end{aligned} \tag{15.27}$$

Again, the (R,S,T) may be functions: their vanishing gives the Maurer-Cartan equation for the orthogonal group $O(n+1,1)$, $\omega^i = 0$ is completely integrable, and its leaves are the fibers of π.

In Part III of the Collected Works of Cartan may be found a wealth - indeed Mother Lode - of raw material for a systematic theory of Connections associates with most of the coset spaces G/K which appear on Klein's "Program." The new feature Cartan added was to study - using his theory of Exterior Differential Systems, created expressly for the purpose - the properties of submanifolds of X and the analytical properties of the "curvature" functions: this is, of course, the "differential" part of "idfferential geometry!" This material has barely begun to understand in terms of contemporary mathematics.

16. RETURN TO GAUGE FIELDS AND K-INVARIANT CONNECTIONS IN PRINCIPAL BUNDLES WITH STRUCTURE GROUPS.

Let (X,Z,π,K) be a principal fiber bundle with K as structure group. Suppose $H: x \to H_x$ is a field of horizontal subspaces, i.e., a connection that is K-invariant. Let $\mathscr{H}$ be the $\mathscr{F}(X)$-module of vector fields V such that:

$$V(x) \in H_x \text{ for all } x \in X \quad .$$

Let $\mathscr{G}(\pi)$ be the set of vector fields V on X which are projectable under π. Let $\mathscr{K}$ be the Lie algebra of K, considered as a Lie algebra of vector fields on X. VE, the vertical vector fields, is isomorphic (as an $\mathscr{F}(X)$-module) to $\mathscr{F}(X) \otimes \mathscr{K}$.

Suppose U is an open subset of X, with a basis

$$(\omega^i,\omega^a) \quad , \; m+1 \leq a,b \leq n \quad , \quad 1 \leq i,j \leq m = \dim Z$$

of one-forms such that the following conditions are satisfied:

The Pfaffian system

$$\omega^i = 0 \tag{16.1}$$

is completely integrable and defines the fiber of π, i.e., the vertical vectors. $\omega^a = 0$ defines the horizontal vectors,

$$d(\omega^a(\mathscr{K})) = 0 \quad . \tag{16.2}$$

As we have seen in Section 13, the following structure equations are then satisfied:

$$d\omega^i = \omega^i_j \wedge \omega^j \tag{16.3}$$

$$d\omega^a = \lambda^a_{bc}\,\omega^k \wedge \omega^c + \Omega^a \quad , \tag{16.4}$$

where (ω^i_j) are one-forms on U, the Ω^a are two-forms and U such that:

$$\mathscr{K} \lrcorner\, \Omega^a = 0 \quad . \tag{16.5}$$

<u>Theorem 16.1</u>. Let (V_a) be the basis of $\mathscr{K}$ such that:

$$\omega^a(V_b) = \delta^a_b \quad .$$

Set:

$$\omega = \omega^a V_a \quad , \tag{16.6}$$

a $\mathscr{K}$-valued one-form on U. Similarly, set

$$\Omega = \Omega^a V_a \quad , \tag{16.7}$$

a $\mathscr{K}$-valued two-form.

<u>Theorem 16.2</u>. ω and Ω are globally defined, independently of the choices of moving frame used to define them. Further,

$$D\omega = \Omega \tag{16.8}$$

where D is the differential operation on Lie algebra valued one-forms defined in Section 14.

<u>Proof</u>. Note that:

$$\omega(\mathscr{H}) \equiv 0 \qquad \omega(V) = V \quad \text{for } V \in \mathscr{K}$$

These two relations show that ω is indeed defined independently of the moving frame. (16.8) is now readily seen to be equivalent to relations (16.4).

We have just seen that there is an elegant "global" structure for K-invariant connections on the principal bundle X. Of course, "physics" takes place on the base space Z. Let us now examine how things look from that point of view.

Let U be an open subset of Z and let

$$\gamma_U : U \to X$$

be a cross-section map. Set:

$$\omega_U = \gamma_U^*(\omega) \tag{16.9}$$

$$\Omega_U = \gamma_U^*(\Omega) \quad , \tag{16.10}$$

where ω_U and Ω_U are $\mathcal{K}$-valued differential forms on U.

Let U' be another open subset Z with a cross-section $\gamma_{U'}$, and $\omega_{U'}, \Omega_{U'}$ defined similarly. Then, if U and U' are non-empty, there is a map

$$\alpha_{UU'} : U \quad U' \to K \tag{16.11}$$

such that:

$$\gamma_U = \alpha_{UU'} \gamma_{U'} \quad . \tag{16.12}$$

When the U,U' are chosen as open sets from an appropriate open covering of X, the system

$$\{\alpha_{UU'}\}$$

defines the cocycle condition needed to define a *principal bundle* in the sense of Steenrod [86]. It is readily seen that this bundle is isomorphic to the one with which we began. The transformation law for the $\omega_{U'}$ in terms of the ω_U and $\alpha_{UU'}$ are readily derived from (16.8), (16.11) and (16.12). It is non-tensorial, i.e., involves the derivatives of the $\gamma_{UU'}$.

Consider again the structure equations (16.3)-(16.4). Usually, in differential geometry, the (ω^i, ω^a) are given, and the curvature forms Ω^a are to be computed from (16.5), and geometric properties (global or local) are to be deduced. In physics, the emphasis is different; we want to find connections whose curvature satisfies differential equations. Thus, the first step in studying connections from the physical point of view is to derive the differential relations which the connection form ω^a and curvature Ω^a satisfy *automatically* as a consequence of the structure equations. These relations can be derived by applying the exterior derivative to both sides of (16.3):

$$0 = 2\lambda^a_{bc} d\omega^b \wedge \omega^c + d\Omega^a = 2\lambda^a_{bc}(\lambda^b_{de}\omega^d \wedge \omega^e + \Omega^b) \wedge \omega^c + d\Omega^a \quad ,$$

whence:

$$d\Omega^a + 2\lambda^a_{bc}\Omega^b \wedge \omega^c = 0 \quad . \tag{16.13}$$

Equations (16.13) are called the *Bianchi identities*. (16.3)-(16.4) may be thought of as a set of differential equations for (ω^a, Ω^a). Very little is actually known about the set of solutions, except in the simplest case. For example, if:

$$\lambda^a_{bc} = 0 \quad . \tag{16.14}$$

Relations (16.14) are equivalent to the condition that

$$[\mathcal{K},\mathcal{K}] = 0 \quad , \tag{16.15}$$

i.e., $\mathcal{K}$ is an abelian Lie group. In this case (which is, of course, the relevant one for Maxwell theory!), Equations (16.13) become decoupled:

$$\begin{aligned} d\omega^a &= \Omega^a \\ d\Omega^a &= 0 \end{aligned} \tag{16.16}$$

In this case, the set of solutions is governed, locally, by the Poincaré lemma, and globally, by de Rham's theorem. We will go into this case in more detail in Section 17.

The next interesting case is that where $\mathcal{K}$ is a non-abelian Lie group, e.g., SU(2) or SU(3). Unfortunately, the generalization of the Poincaré lemma and de Rham's theorem does not yet exist for this case.

17. THE RELATION BETWEEN CURVATURE AND PRINCIPAL BUNDLES FOR ABELIAN LIE GROUPS

In case $[\mathcal{K},\mathcal{K}] = 0$ the curvature forms Ω^a satisfy

$$d\Omega^a = 0 \quad , \tag{17.1}$$

Together with the condition

$$\mathcal{K} \lrcorner \Omega^a = 0 \quad , \tag{17.2}$$

this implies that there are two-forms F^a on Z such that:

$$dF^a = 0 \tag{17.3}$$

$$\pi^*(F^a) = \Omega^a = d\omega^a \quad . \tag{17.4}$$

Conversely, given a set of two-forms

$$F^{m+1},\ldots,F^n \tag{17.5}$$

on a manifold Z, satisfying

$$dF^a = 0 \quad , \tag{17.6}$$

we ask: When are the curvature forms on a K-invariant connection in a principal bundle with structure group K? The answer is well known, involving a bit of elementary fiber bundle topology, and has played an important role for thirty years in algebraic geometry, quantum mechanics ("geometric quantization"), and Lie group harmonic analysis (Kirillov-Kostant-Auslander theory of unitary representations of nilpotent and solvable Lie groups). It also turns out to be the appropriate geometric formalism to express Dirac's "quantization" condition for magnetic monopoles.

Write the F's satisfying (17.5) vectorially:

$$F = \begin{pmatrix} F^{m+1} \\ \vdots \\ F^n \end{pmatrix} \tag{17.7}$$

Regard F as a two-differential form with values in the vector space R^{n-m}.

By the Poincaré lemma, Z can be covered with open subsets $\{U\}$ in each of which there is a matrix of one-forms θ_U such that:

$$d\theta_U = F \quad . \tag{17.8}$$

Whenever U and U' are two elements of the open covering which intersect, in U U', we have:

$$d\theta_U = F = d\theta_{U'} \quad ,$$

i.e.,

$$d(\theta_U - \theta_{U'}) = 0 \quad . \tag{17.9}$$

Now, it is possible to choose the covering $\{U\}$ so that it has the following property:

> Whenever a finite number of open subsets of the covering intersect, its intersection is diffeomorphic to R^m. (17.10)

(For example, one can choose the U to be *convex* with respect to an affine connection of Z, by a theorem of J.H.C. Whitehead.)

Since $U \cap U'$ is R^n, we can apply the Poincaré lemma again, and obtain a vector-valued function $f_{UU'}$ such that:

$$df_{UU'} = \theta_U - \theta_{U'} \quad . \tag{17.11}$$

In the intersection of *three* open sets of the covering, U, U', U", we have:

$$df_{UU'} + df_{UU'U"} + df_{U"U} = 0$$

or

$$f_{UU'} + f_{U'U"} + f_{U"U} = \text{constant, say } v_{UU'U"} \in R^{n-m} \tag{17.12}$$

This assignment

$$(U,U',U") \quad v_{UU'U"} \tag{17.13}$$

defines a two-cochain in the Cech cohomology of Z (with coefficients in R^{n-m}) with respect to this covering. From (17.12), we see that this cochain is a cycycle. A theorem of A. Weil then determines an element of

$$H^2(Z,R^{n-m})$$

the second cohomology group of Z with coefficients in R^{n-m}. Since $\mathcal{K}$ is isomorphic (as a vector space) to R^{n-m}, the Čech cocycle $\{v_{UU'U"}\}$ determines an element of

$$H^2(Z,\mathcal{K}) \quad , \tag{17.14}$$

the vector space of the second Čech cohomology group of the *manifold* Z with coefficients in the *vector space* $\mathcal{K}$. (Of course, this is isomorphic to the de Rham cohomology.) (I apology to topologists for the choice of notation. In topology, Z usually denotes integers.)

Remark. For more detail about this sort of argument, see the excellent treatment by V. Guillemin, S. Sternberg and N. Wallach [27,89] These treatises also contain much more detail about the relation to Lie group harmonic analysis and quantum mechanics. This argument is basically due to Weil.

In each U,U' such that $U \cap U"$ is non-empty, construct the map

$$\phi_{UU'}: U \cap U' \to \mathcal{K} \tag{17.15}$$

as follows:

$$\phi_{UU'}(z) = f^1_{UU'} V_1 + \dots + f^{n-m}_{UU'} V_{n-m} \tag{17.16}$$

(Recall that $(V_1,\dots,V_{n-m})$ is a fixed basis of $\mathcal{K}$, the Lie algebra of the connected abelian Lie group K.)

Since the Lie group K is abelian, the map

$$\exp: \mathcal{K} \to K$$

is onto and a local diffeomorphism. For each (U,U') set

$$k_{UU'} = \exp(\phi_{UU'}) \quad . \tag{17.17}$$

The central question is now:

Can the system of the $\{k_{UU'}\}$ be chosen so that whenever U, U', U'' are three elements of the covering which intersect, the following condition is satisfied: (17.18)

$$k_{UU'} = k_{UU''} k_{U''U'} \quad .$$

If (17.18) is satisfied, then there is a principal bundle with structure group K, by a standard construction. It is then readily verified that the $\{\theta_U\}$ considered as a differential form with value in $\mathcal{K}$, defines a connection for this bundle, for which $F^1, \ldots, F^{n-m}$ are the curvature forms. The analysis of the condition that (17.18) be satisfied is now straightforward, but length I will only state the result in the following form:

<u>Theorem 17.1</u>. Let K be a connected abelian Lie group with Lie algebra $\mathcal{K}$. Let $\pi_1(K)$ be its fundamental group. Regard $\pi_1(K)$ as a subgroup of the translation group of (as the kernel of the exponential map $\mathcal{K} \to K$). Let F be a $\mathcal{K}$-valued two-form on a manifold Z, such that

$$dF = 0 \quad .$$

(Of course, DF also equals zero, since $\mathcal{K}$ is an abelian Lie algebra.) Via the de Rham-Cech-Weil isomorphism, F determines an element ch(F) of the cohomology group

$$H^2(Z, \mathcal{K})$$

of Z with coefficients in the vector space $\mathcal{K}$. Then, a necessary and sufficient condition that F be the curvature tensor of an invariant connection on a principal bundle X with structure group K is that:

$$ch(F) \in H^2(Z, \pi_1(K)) \quad , \tag{17.19}$$

where $\pi_1(K)$ is regarded as a subgroup of the abelian translation group of the vector space $\mathcal{K}$.

Thus, we see that there are two obvious sufficient conditions that *every* closed $\mathcal{K}$-valued two-form be a curvature tensor:

$$\pi_1(K) = 0 \tag{17.20}$$

i.e., K is simply connected, or

The second cohomology groups of Z with coefficients in the integers is zero. (17.21)

Looked at from another point of view, if neither (17.20) nor (17.21) is satisfied, the condition (17.9) will give us a sort of *quantization condition* for the "gauge field" F. Thus, it is of physical interest to start with Z's and K's for which $H^2(Z, \pi_1(K))$ can be nonzero. For example, an interesting choice is

$$K = U(1) = SO(2, R)$$

$$Z = R^3 - \text{(one point)} \quad .$$

The "quantization condition" (17.19) is, in this case, closely related to Dirac's *magnetic monopole quantization condition.*

18. MAXWELL'S EQUATIONS

We have seen that interpreting a finite set of two-differential forms as an abelian Lie algebra-valued differential forms leads to interesting geometry. In particular, *topology* enters for the first time.

The place in physics to look for two-differential forms is the *Maxwell theory of electromagnetism.* We will begin with them in the traditional vector analysis notation. (The treatises by Jackson [71], Born and Wolf [3] are now the standard references in the physics literature.)

$$\frac{1}{c}\frac{\partial \vec{B}}{\partial t} = -\operatorname{curl}\vec{E} \tag{18.1}$$

$$\frac{1}{c}\frac{\partial \vec{D}}{\partial t} + \frac{4\pi}{c}\vec{J} = \operatorname{curl}\vec{H} \tag{18.2}$$

$$\operatorname{div}\vec{B} = 0 \tag{18.3}$$

$$\operatorname{div}\vec{D} = 4 \tag{18.4}$$

$$\frac{\partial\rho}{\partial t} + \operatorname{div}\vec{J} = 0 \quad . \tag{18.5}$$

Here, all quantities, except charge density ρ, which is a scalar, are time-dependent vector fields on R^3 (i.e., maps $R^3 \times R \to R$).

$\vec{E}$ = electric field , $\vec{H}$ = magnetic field

$\vec{B}$ = magnetic induction , $\vec{D}$ = electric induction

$\vec{J}$ = current , ρ = charge density

c = a constant, which turns out to be the velocity of light

Now, (18.1)-(18.5) constitute four differential equations for the six unknown functions listed above. (Equation (18.5) is a consequence of the other equations.) Other equations are needed to make a determined problem; they are called *constitutive relations*, and are usually prescribed by the physical situation. Of course, "c" is *velocity of light,* (presumably) a constant!

As the first step in bringing the geometric structure of electromagnetics to the foreground, let us rewrite Maxwell's equation (18.1)-(18.5) in terms of differential forms on R^3 and the "natural" operators on differential forms (exterior derivative, exterior multiplication, etc.). Recall that

$$d: \mathcal{D}^r(R^3) \to \mathcal{D}^{r+1}(R^3);\ r = 0,1,2,\ldots$$

denotes *exterior derivative* of differential forms.

$d: \mathcal{D}^1(R^3) \to \mathcal{D}^2(R^3)$ is essentially *curl*, while $d: \mathcal{D}^2(R^3) \to \mathcal{D}^3(R^3)$ is essentially *div*. Looking at Maxwell's equation, this tells us how the data is to be interpreted in terms of differential forms:

$\vec{D},\vec{B}$ are time-dependent elements of $\mathcal{D}^2(R^3)$

$\vec{E},\vec{H}$ are time-dependent elements of $\mathcal{D}^2(R^3)$

ρ is a three-form, $\vec{J}$ a two-form.

In order to indicate to the reader that we are using these conventions, we will *leave off these arrows.* Thus, we have:

$$\frac{1}{c}\frac{\partial B}{\partial t} = - dE \qquad (18.6)$$

$$\frac{1}{c}\frac{\partial D}{\partial t} + \frac{4\pi}{c} J = dH \qquad (18.7)$$

$$dB = 0 \qquad (18.8)$$

$$dD = 4\pi\rho \qquad (18.9)$$

$$\frac{\partial \rho}{\partial t} + dJ = 0 \quad . \qquad (18.10)$$

It is well-known that these equations can be written even more elegantly and compactly in terms of differential forms on R^4. However, in the classical theory of electromagnetics and optics, time *does* play a special rôle, and these equations are usually more convenient for calculations.

It is traditional to rewrite Maxwell's equations as integral relations and to derive certain "shock-wave" conditions in the case that certain data develop singularities. One immediate (and even decisive) payoff from the differential form reformulation is that this process is remarkably simple. (Of course, differential forms were *invented* precisely in order to handle "integration" in a smooth and coordinate-free way. The "right" mathematical setting for these ideas is in *de Rham's Theory of Currents* [82]. It is also related to the differential-geometric theory of characteristics and shock waves developed in *Differential Geometry and the Calculus of Variations* and *Geometry, Physics and Systems*, Interdisciplinary Mathematics, Volumes 5 and 9.)

Equations (18.6) and (18.7) are equations in $\mathcal{D}^2(R^3)$. In order that they make sense, the data must be at least differentiable. We would like to convert them into equations which would allow singularities, e.g., discontinuities. In order to do this, exterior multiply both sides with a one-form θ *of compact support* (i.e., θ vanishes outside of a compact subset of R^3), and then integrate over R^3.

$$\int_{R^3} \frac{1}{c}\theta \wedge \frac{\partial B}{\partial t} = - \int_{R^3} \theta \wedge dE \qquad (18.11)$$

$$\int_{R^3} \frac{1}{c}\theta \wedge \frac{\partial D}{\partial t} + \frac{4\pi}{c}\theta \wedge J = \int_{R^3} \theta \wedge dH \qquad (18.12)$$

Use the algebraic rules relating exterior derivation and multiplciation:

$$\theta \wedge dH = - d(\theta\ H) + d\theta \wedge H \quad .$$

Hence, using Stokes' formula and the assumption that θ vanishes at infinity on R^3,

$$\int_{R^3} \theta \wedge dH = \int_{R^3} d\theta \wedge H \quad .$$

Thus, we have proved:

<u>Theorem 18.1</u>. If B, E, D and H are C^1 solutions of the Maxwell equations (18.6)-(18.7) then these equations are equivalent to the following integro-differential equations:

$$\frac{1}{c}\frac{d}{dt}\int_{R^3} \theta \wedge B = - \int_{R^3} d\theta \wedge E \qquad (18.13)$$

The right hand side of (18.15) can now be converted to an integral over $S_{t,\varepsilon}$, $S_{t,-\varepsilon}$ using Stokes' formula. First,

$$d\theta \wedge E = d(\theta \wedge E) + \theta \wedge dE \tag{18.16}$$

in $R^3 - T_{t,\varepsilon}$. Substitute (18.16) into (18.15):

$$\frac{d}{dt}\frac{1}{c}\int_{R^3 - T_{t,\varepsilon}} \theta \wedge B = -\int_{R^3 - T_{t,\varepsilon}} \theta \wedge dE - \int_{S_{t,-\varepsilon}} \theta \wedge E - \int_{S_{t,\varepsilon}} \theta \wedge E. \tag{18.17}$$

I do not want to go into the full details here of the analysis required to deduce the full consequences of these relations. At the intuitive level customarily used in physics books, let us suppose that E and B are "smooth" off of S_t, but have a "jump" discontinuity across S_t. The most plausible way to satisfy (18.17) is to require that

$$\frac{1}{c}\frac{\partial B}{\partial t} = -dE$$

in $R_3 - S_t$, i.e., Maxwell's *differential* equations are satisfied off of the surface S_t, and that the second and third terms on the right hand side of (18.17) vanish. Now, $S_{t,-\varepsilon}$, $S_{t,\varepsilon}$ (recall that they are the boundaries of the ε-tubular neighborhood of S_t) are *oppositely* oriented. Thus, it is appropriate to deduce from (18.17), as $\varepsilon \to 0$, the following relation:

For each point $x \in S_t$, each $t \in R$, let E_+, E_- be the covectors at x obtained as the limiting value of the covectors $E(y)$, as $y \to x$ from one side and the other of the surface S_t.

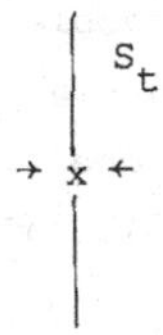

Thus,

$$(E_+ - E_-)(v) = 0 \tag{18.18}$$

for each tangent vector v to the surface S_t.

This sort of reasoning (which the specialists in partial differential equations have completely mechanized) leads to the following sort of statement:

The integrated form of Maxwell's equations, applied to fields with simple "jump" discontinuities of E and H, leads to the further condition that the "tangential components" of E and H are *continuous* across the surfaces of discontinuity, i.e., the "wave fronts."

So far, we see no conditions on the surfaces of discontinuity. This is because the equations are "underdetermined." By imposing "constitutive relations," further conditions are obtained--the surfaces S_t satisfy a first order partial differential equation--the *characteristic equations*--and the *normal components* of E and H satisfy *transport equations*.

$$\frac{1}{c}\frac{d}{dt}\int_{R^3}\theta \wedge D + \frac{4\pi}{c}\int_{R^3}\theta \wedge J = \int_{R^3} d\theta \wedge H \tag{18.14}$$

for each compactly supported one-form θ on R^3.

Thus, the first two of the Maxwell equations have been transformed into a form *so that they make sense for non-smooth objects* (or, at least, objects which have a lesser degree of smoothness, e.g., continuous, but piecewise C^∞). The equations (18.8)-(18.9) can be transformed in a similar way.

Let us further illustrate the utility of the differential form formalism (and the freedom it gives one to work and think in a coordinate-free way) by deriving from (18.3)-(18.4) the conditions which must be satisfied for a time-dependent surface on R^3 to carry a singularity (e.g., a "shock wave" or a "wave front") for a Maxwell field. Again, this is a traditional topic, which is extremely important for *geometric optics*. Again, this is a rather beautiful classical idea that has been abstracted and vastly extended in the modern partial differential theory literature - "Fourier integral operators," "wave front sets," etc. The book by Guillemin and Sternberg, *Geometric Asymptotics* [27], gives an excellent treatment of some of this material, although much remains to be done to fully elucidate the complicated and fascinating inter-relations between the mathematics and physics.)

Let us consider Equations (18.13)-(18.14), which hold for *each* compactly supported one-form , but let us *not* suppose that B and E are C^1, so that Stokes' formula cannot be applied in a convenient way to deduce (18.6). Of course, in an open subset of R^3 in which B and E are C^1, θ can be chosen to vanish outside of that subset, Stokes' formula can be applied, and we then deduce (18.6) *on that open set*. Thus, we have to deduce the consequences of (18.13) at points on which B and E are not C^1.

Let us make the following assumption

For each t, S_t is a surface in R^3. B and E are C^1 in $R^3 - S_t$.

Let be a positive small number and let $S_{t,\varepsilon}$, $S_{t,-\varepsilon}$ be the boundary of a tubular neighborhood surrounding S_t:

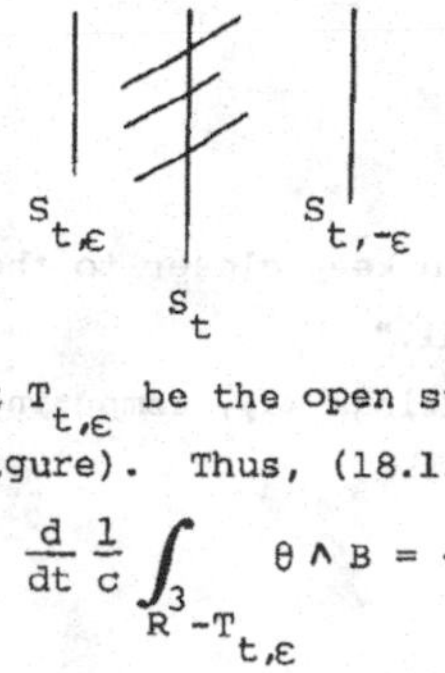

Let $T_{t,\varepsilon}$ be the open subset of R^3 between $S_{t,\varepsilon}$ and $S_{t,-\varepsilon}$ (the shaded area in the above figure). Thus, (18.13) implies that

$$\frac{d}{dt}\frac{1}{c}\int_{R^3-T_{t,\varepsilon}} \theta \wedge B = -\int_{R^3-T_{t,\varepsilon}} d\theta \wedge E \quad . \tag{18.15}$$

Let us now turn to the more esthetically satisfying (at least to a mathematician!) four dimensional theory, which takes place on a manifold Z that has at least some of the attributes of our usual "space-time."

Let a Riemannian metric be given on Z as a symmetric, non-degenerate $\mathcal{F}(Z)$-bilinear map

$$<,>: (V_1, V_2) \to <V_1, V_2>$$

on $\mathcal{V}(Z)$. This bilinear form dualizes to define another, that we also denote by $<,>$, on the one-differential forms $\mathcal{D}^1(Z)$. It then extends to define a bilinear form

$$<,>: \mathcal{D}^r(Z) \times \mathcal{D}^r(Z) \to \mathcal{F}(Z)$$

on the r-differential forms on Z.

Suppose the manifold Z is connected and orientable such taht dim Z = n. Recall that an *orientation* for Z is an equivalence class of everywhere non-zero n-forms on Z, where one says that $\omega, \omega' \in \mathcal{D}^n(Z)$ are *equivalent* if

$$\omega = f\omega'$$

with $f \in \mathcal{F}(Z)$, $f(z) > 0$ for all $z \in Z$. Suppose one orientation is chosen for Z. We say that a form $\omega \in \mathcal{D}^n(Z)$ is a *volume element form* for Z if it belongs to the chosen orientation class, and if $(z) \neq 0$ for all $z \in Z$.

Suppose a Riemannian metric is given. One proves that there is a unique volume element form ω such that $<\omega,\omega>^2 = 1$. We will use this form ω to define integrals over Z

$$\int_Z f\omega$$

$$f \in \mathcal{F}(Z) \quad .$$

It is convenient to write

$$\omega = dz \quad ,$$

so that the integral can be written in a form closer to the usual calculus notation:

$$\int_Z f(z)\,dz \quad .$$

Let $\mathcal{D}_0^1(Z)$ denote the one-forms with *compact support*, i.e., those one-forms which vanish outside of some compact subset of Z. Define a real-valued function

$$L: \mathcal{D}_0^1(Z) \to R$$

by the following formula:

$$L(A) = \int_Z <dA,dA>\,dz \quad \text{for } A \in \mathcal{D}_0^1(Z) \quad . \tag{18.19}$$

<u>Remark</u>. I now use the notation "A" for elements of $\mathcal{D}^1(Z)$ to keep closer to the physics notation, where "A" stands for "electromagnetic potential."

Let us extremize (18.19) in the usual calculus of variations way, computing the *first variation*

$$dL(A,\delta A) \equiv \frac{d}{d\varepsilon} L(A+\varepsilon\delta A)\Big|_{\varepsilon=0} \quad , \tag{18.20}$$

where δA is another element of $\mathcal{D}_0^1(Z)$. In fact,

$$dL(A,\delta A) = 2\int_Z \langle d\delta A,\ dA\rangle\ dz \quad . \tag{18.21}$$

Let us pause in the main development to recall another topic, the Hodge *-operation. It is an $\mathcal{F}(Z)$-linear map $\mathcal{D}^r(Z) \to \mathcal{D}^{n-r}(Z)$ $r = 0,.,e,...,$ which is essentially defined by the following identity

$$\theta_1 \wedge * \theta_2 = \langle\theta_1,\theta_2\rangle\ dz \quad \text{for } \theta_1,\theta_2 \in \mathcal{D}^r(Z) \ . \tag{18.22}$$

Let us apply (18.22) to the integrand of (18.23), with r = 2:

$$\langle d\delta A,\ dA\rangle\ \ dz = d\delta A \wedge * \ dA \ \ . \tag{18.23}$$

Hence,

$$dL(A,\phi A) = 2\int_Z d\phi A \wedge * \ dA = 2\int_Z d(\delta A \wedge * \ dA) + \delta A \wedge d * dA \quad . \tag{18.24}$$

Using Stoke's theorem and the assumption that dA has compact support, the first term on the right hand side of (18.24) vanishes. We are left with

$$dL(A,\delta A) = 2\int_Z \delta A \wedge d * dA \tag{18.25}$$

Now, in order that $A_0 \in \mathcal{D}_0^1(Z)$, extremize this "action" function $A \to L(A)$, we must have:

$$dL(A_0,\delta A) = 0 \quad \text{for } \mathit{all}\ \delta A \quad . \tag{18.26}$$

In view of (19.25), this forces

$$d{*}dA = 0 \ \ . \tag{18.27}$$

Notice that (18.27) is a set of second order linear partial differential equations for A. However, notice also that these differential equations are of a more sophisticated type than the usual linear partial differential equations of mathematical physics (e.g., Laplace and Poisson equations, wave equations, Helmholtz equations, etc.). In particular, (18.27) has a property physicists call *gauge invariance*. Since dd = 0, we can add to any solution A a one-form θ with $d\theta = 0$ to obtain another solution, $A + \theta$. This property leads to all sorts of complications in the physics and engineering literature. Certain special choices of "gauges" (i.e., *additional* differential equations) have to be imposed so that Equation (18.27) may be replaced by a system of the following form

$$\frac{\partial\psi}{\partial t} = D\psi \ , \tag{18.28}$$

where ψ is a vector-valued function on Z (constructed in some way from A), t is one of the coordinates on Z, and D is a differential operator in a remaining set of coordinates for Z. (Differential equations of type (18.28) are said to be of *evolution type*.) Another alternative for handling Equations (18.27) in a convenient way is to set F = dA, a two-form, and consider (18.27) as replaced by the following differential equations:

$$d{*}F = 0 \quad , \ dF = 0 \tag{18.29}$$

These are the free Maxwell equations themselves.

We can now relate these general Maxwell equations to the physicist's version given in Section 17. Specialize Z to be R^4, space-time, with the Lorentz metric. Label Cartesian coordinates on Z as

$$(t,x^i) \quad , \quad 1 \leq i,j \leq 3 \ .$$

The Lorentz metric is then

$$ds^2 = dt^2 - \delta_{ij} dx^i \quad dx^j \quad . \tag{18.30}$$

δ_{ij} is the "Kronecker delta" matrix, i.e., zero if $i \neq j$, equal to one if $i = j$. (We have normalized the velocity of light to be one.)

Let

$$\left(\frac{\partial}{\partial t}, \frac{\partial}{\partial_x i}\right)$$

be the basis for vector fields on R^4 dual to these coordinates. In the metric (18.30), the inner product $< , >$ on vector fields takes the following form:

$$< \frac{\partial}{\partial t}, \frac{\partial}{\partial t} > = 1$$

$$< \frac{\partial}{\partial_x i}, \frac{\partial}{\partial_x j} > = - \delta_{ij}$$

$$< \frac{\partial}{\partial t}, \frac{\partial}{\partial_x i} > = 0 \quad .$$

One derives the following formula for inner product of differential one-forms:

$$< dt, dt > = 1$$

$$< dx^i, dx^j > = - \delta^{ij}$$

$$< dx^i, dt > = 0 \ .$$

Here,

$$dz \equiv dx^1 \wedge dx^2 \wedge dx^3 \wedge dt$$

is the normalized volume element differential form.

If F is a two-form on Z, it can be written in a unique way as

$$F = B + E \wedge dt \quad . \tag{18.31}$$

B is a two-form in dx^i, E a one-form in dx^i with coefficients that depend on x and t. Now,

$$dB = dt \wedge \frac{\partial B}{\partial t} + d_x B \tag{18.32}$$

where

$$\frac{\partial B}{\partial t} \equiv \mathscr{L}_{\frac{\partial}{\partial t}} (B)$$

is the two-form resulting from differentiating the coefficients with respect to t. $d_x B$ is the exterior derivative of B as differential form on x if the variable is held free

$$dF = dt \wedge \frac{\partial B}{\partial t} + d_x B + d_x E \wedge dt \quad . \tag{18.33}$$

The first Maxwell equation, $dF = 0$, then takes the form

$$\frac{\partial B}{\partial t} + d_x E = 0 \qquad\qquad d_x B = 0 \quad . \tag{18.34}$$

We must now calculate *F. It is a two-form, that can also be written in the form (18.33):

$$*F = B' + E' \wedge dt \quad .$$

Let $F_1 = B_1 + E_1$ dt be an arbitrary two-form. Then,

$$\langle F,F_1\rangle = \langle B,B_1\rangle + \langle E \wedge dt,\ E_1 \wedge dt\rangle$$

$$= \langle B,B_1\rangle + \langle E,E_1\rangle$$

$$F_1 \wedge *F = (B_1 + E_1 \wedge dt) \wedge (B' + E' \wedge dt)$$

$$= (E_1 \wedge B') \wedge dt + (B_1 \wedge D') \wedge dt \quad .$$

Hence,

$$F_1 \wedge *F = \langle F,F_1\rangle \wedge dx \wedge dt$$

with

$$dx = dx^1 \wedge dx^2 \wedge dx^3 \quad .$$

Thus,

$$E_1 \wedge B' + B_1 \wedge E' = (\langle B,B_1\rangle + \langle E,E_1\rangle) \quad dx \qquad (18.35)$$

We now want to find formulas for *F from this identity. First, set

$$E_1 = 0 \quad ,$$

resulting in the following:

$$B_1 \wedge E' = \alpha\langle B,B_1\rangle\ dx$$

or

$$E' = - *_3 B \quad , \qquad (18.36)$$

where $*_3$ denotes the Hodge dual operator on R^3 with respect to the (positive) Euclidean metric on R^3. Now set $B_1 = 0$ and obtain

$$E_1 \wedge B' = \langle E,E_1\rangle\ dx \quad ,$$

or

$$B' = - *_3 E \quad .$$

Thus,

$$- *F = *_3 E + *_3 B \wedge dt \quad . \qquad (18.37)$$

The second half of Maxwell's equation, d*F = 0, now takes the following form:

$$d_*(*_3 E) = 0 \qquad \partial_t(*_3 E) + d_*(*_3 B) = 0 \quad . \qquad (18.38)$$

Thus, Equations (18.34) and (18.38) are essentially the free Maxwell equations (18.6)-(18.9) (with c = 1, E = D, B = H, $\rho = 0$).

$$E \to d_x *_3 E \quad , \quad B \to d_x B$$

are essentially the *divergence* equations of classical vector analysis, E → dE, and B → d_xB are the *curl*. (Notice that classical vector analysis identifies one-forms and two-forms on R^3 via the Hodge duality operator $*_3$.)

After this review of the classical theory, let us turn to a treatment that is moro in tune with modern differential geometry.

19. MAXWELL THEORY ON A GENERAL 4-DIMENSIONAL MANIFOLD

Let Z be a *four*-dimensional manifold. Let $\mathcal{D}^2(Z)$ be the (Z)-module of two-differential forms on Z. Suppose given a map (not necessarily linear)

$$CR: \mathcal{D}^2(Z) \to \mathcal{D}^2(Z) \quad .$$

("CR" stands for "constitutive relation.") Let

$$J, J' \in \mathcal{D}^3(Z) \quad .$$

Maxwell's equations are then the differential equations for a pair

$$(F,F) \in \mathcal{D}^2(Z) \times \mathcal{D}^2(Z)$$

of two-forms on Z such that:

$$dF = J'; \qquad dF' = J'; \qquad Z' = CR(Z) \quad . \tag{19.1}$$

In the usual electromagnetic theory, without "magnetic poles," one also assumes:

$$J' = 0 \quad . \tag{19.2}$$

The form of distribution theory created by G de Rham [82] ("the theory of currents") is obviously relevant here. (Of course, the name "current" indicates that de Rham had electromagnetic theory in mind!) Thus, one can ask for solutions of (19.1)-(19.3) with (J,J') in a space of such currents. Let supp(J,J') be the support, a closed subset of Z. Then, in

$$Z - \mathrm{supp}(J,J') \quad ,$$

(F,F') will be the *closed* two-forms, and--if they are, in addition, C^∞--bundles can be constructed, as in Section 18. This indicates a whole class of problems of a mixed analytic and geometric nature, which have barely begun to be explored.

Yang-Mills Theory is a natural generalization. Choose a Lie group K. Construct principal bundles with K as structure group and Z (or certain submanifolds of Z) as base, with K-invariant connections, *subject to certain equations on the curvature.* Analytically, this involves replacing (19.1)-(19.2) with a set of equations involving Lie algebra-valued differential forms. What the physicists call "Yang-Mills fields" are quantum structures of this type. Of course, Einstein's theory of gravitation (and Newton's, in the form it was put by Cartan [8]) is also of this form.

In the underlying physical theories, the three-forms (J,J') are to be investigated further. Often they are constructed from cross-sections of other bundles on Z, which physically represent additional fields. The differential equations for these fields, plut (19.1)-(19.3), then represent a set of "interacting fields." The study of these "interacting field equations" and the associated geometric analytic-algebraic problems is the essence of mathematical physics.

Let "t" be an element of $\mathcal{F}(Z)$, i.e., real-valued function on Z. Suppose that

$$dt \neq 0 \quad \text{at each point of } Z.$$

Each two-form, F on Z can then be written (perhaps only locally) in the form

$$F = E \wedge dt + B$$

where E is a one-form, B a two-form, such that the forms E,B do not contain "dt," but may contain "t" as a parameter in the coefficients. Explicitly, if (x^i), $1 \leq i,j \leq 3$

are functions on Z such that (t,x^i) is a coordinate system, then

$$E = E_i dx^i \quad , \quad B = \varepsilon_{ijk} B^i dx^j \wedge dx^k$$

$$dF = d'E \wedge dt + d'B + B_t \wedge dt$$

where d' is the exterior derivative of these forms *with* t *held fixed* and B_t is the partial derivative with respect to t (or the Lie derivative with respect to $\partial/\partial t$).

Now, physicists need more than just the bare bones of a set of differential equations. Of course, precisely what it is that they want is part of the mystery (to mathematicians) of physics! Certainly, we all know (especially these days) that they need to know what "energy" is. Of course, classical Maxwell theory has an expecially beautiful answer here--*the electromagnetic stress-energy tensor of the electromagnetic field.* I will now briefly present the formula for it presented in "Interdisciplinary Mathematics," Volume 4, page 72. This involves using a Riemannian metric for Z to define the map "CR." Recall that a *Riemannian metric* for Z is an $\mathcal{F}(Z)$-bilinear, symmetric map

$$\mathcal{V}(Z) \times \mathcal{V}(Z) \to \mathcal{F}(Z)$$

$$(V_1,V_2) \to \langle V_1,V_2\rangle \quad ,$$

such that

$$\langle V, \mathcal{V}(Z)\rangle = 0$$

implies $V = 0$

i.e., the field of symmetric, bilinear forms determined on the tangent spaces to Z are non-degenerate.

One can then extend this form to define bilinear forms on all the tensor fields; in particular, to $\mathcal{D}^r(X)$:

$$\theta_1,\theta_2 \to \langle\theta_1,\theta_2\rangle \ .$$

Assume X is oriented as a manifold. The *volume element form* is the unique *positively oriented* four-form dz such that

$$\langle dx, dx\rangle^2 = 1 \quad .$$

(Notice that we are not assuming the metric to be positive.) The *Hodge dual* is then an $\mathcal{F}(Z)$-linear mapping

$$\mathcal{D}^r(Z) \to \mathcal{D}^{4-r}(Z) \quad ,$$

$$r = 1,2,3 \quad ,$$

$$\omega \to *\omega \quad ,$$

such that:

$$\theta \quad \omega = \langle\theta,*\omega\rangle \, dz \qquad \text{for } \theta \in \mathcal{D}^r(Z), \quad \omega \in \mathcal{D}^{4-r}(Z) \tag{19.3}$$

<u>Theorem 19.1</u>. For $\theta \in \mathcal{D}^1(Z)$, $\omega \in \mathcal{D}^r(Z)$,

$$*(\omega \wedge \theta) = V \lrcorner *\omega \tag{19.4}$$

where V is the vector field on Z such that:

$$\theta'(V) = \langle\theta,\theta'\rangle \qquad \text{for } \theta' \in \mathcal{D}^1(Z). \quad . \tag{19.5}$$

(In words, V is the image in $\mathscr{V}(Z)$ of θ under the module isomorphism $\mathscr{D}^1(Z) \to \mathscr{V}(Z)$ induced by the metric.)

Proof. We will only prove this in the case <,> is positive. (The general case is just a matter of putting $\sqrt{-1}$ in the right places!) It suffices to prove (19.4) in case ω and θ are normalized to:

$$<\omega,\omega> = 1 \quad , \quad <\theta,\theta> = 1 \quad , \tag{19.6}$$

$$\omega = \theta_1 \wedge \ldots \wedge \theta_r \quad , \tag{19.7}$$

where $\theta_1,\ldots,\theta_r \in \mathscr{D}^1(Z)$, $<\theta_i,\theta_j> = \delta_{ij}$ for $1 \leq i,j \leq r$.

Case 1. $\theta = \theta_1$, $*\omega = \theta_{r+1} \wedge \ldots \wedge \theta_4$ where $(\theta_1,\ldots,\theta_r)$ are an orthonormal basis. Both sides of (19.6) are zero.

Case 2. $\theta = \theta_{r+1}$. In this case,

$$\omega \quad \theta = \theta_1 \wedge \ldots \wedge \theta_{r+1} \qquad *(\omega \wedge \theta) = \theta_{r+2} \wedge \ldots \wedge \theta_4 \tag{19.8}$$

$$V \quad *\omega = \theta_{r+1}(V)\ \theta_{r+2} \wedge \ldots \wedge \theta_4 - \theta_{r+1} \wedge \theta_{r+2}(V) \quad \ldots + \ldots$$

$$= \theta_{r+1}(V)\theta_{r+2} \wedge \ldots \quad \theta_4 + \underline{\underline{0}}$$

$$= \theta_{r+2} \wedge \ldots \wedge \theta_4 \tag{19.9}$$

Equality of (19.8) and (19.9) finishes the proof.

The general case follows from these two cases.

Formula (19.4) is the basic algebraic lemma in the Maxwell theory from this point of view. Let us consider Maxwell's equations in the following form:

$$dF = 0 \quad , \quad d * F = J \quad . \tag{19.10}$$

Suppose we write

$$F = E \wedge dt + B \tag{19.11}$$

where "t" is a function on Z such that

$$<dt,dt> = -\frac{1}{c^2} \quad , \quad c \text{ a constant.} \tag{19.12}$$

Suppose ∂_t is the vector field image of dt in $\mathscr{V}(Z)$. Then, $*F = \partial_t \lrcorner *E + *B$.

Suppose D is the form which does not contain dt, such that

$$*E = D \wedge dt \quad . \tag{19.13}$$

Suppose H does not contain dt and:

$$*B = H \wedge dt \quad . \tag{19.14}$$

Then,

$$*F = \partial_t \lrcorner (D \wedge dt) + H \wedge dt$$

$$= -D + H \wedge dt \tag{19.15}$$

$$d * F = -\mathscr{L}_{\partial_t}(-D) \wedge dt * - d'D + d'H \wedge dt$$

$$= \mathscr{L}_{\partial_t}(D) \wedge dt - d'D + d'H \wedge dt \tag{19.16}$$

Also,

$$dF = -\mathcal{L}_{\partial_t}(E) \wedge dt + d'E - d'B \wedge dt \tag{19.17}$$

Suppose

$$J = j + \rho \wedge dt \tag{19.18}$$

Thus, the Maxwell equations are:

$$\mathcal{L}_{\partial_t}(E) = d'B \tag{19.19}$$

$$d'E = 0 \tag{19.20}$$

$$\mathcal{L}_{\partial_t}(D) = -d'H + \rho \tag{19.21}$$

$$d'D = -j \quad . \tag{19.22}$$

20. ENERGY IN THE MAXWELL THEORY

Keep the notation of Section 19, with the Maxwell theory based on a Hodge dual operator $*: \mathcal{D}^r(F) \to \mathcal{D}^{4-r}(Z)$ associated with a Riemannian metric $\langle , \rangle$ on Z. Denote the associated volume element form "dz."

<u>Definition</u>. Let F be a two-form on Z. Define

$$T: \mathcal{V}(Z) \times \mathcal{V}(Z) \to \mathcal{F}(Z)$$

by the following formula:

$$T(V_1, V_2) = \frac{1}{2}\langle V_1, V_2\rangle\langle F, F\rangle - \langle V_1 \lrcorner F, V_2 \lrcorner F\rangle \tag{20.1}$$

T is then an $\mathcal{F}(Z)$-bilinear map (hence, defines a twice covariant tensor field, i.e., a cross-section of $T^d(Z) \circ T^d(Z)$, the symmetric tensor product of the cotangent bundle). It is called the *energy momentum tensor* of the electromagnetic field associated with the two-form F.

<u>Remark</u>. Barut's book [1] has, in my opinion, the best treatment of this topic in the physicist's style.

<u>Remark</u>. This is the "symmetric" energy-momentum tensor (which is not in general unique), defined via the Belinfante-Rosenfeld method, i.e., involving a "deformation" of the metric $\langle , \rangle$ as explained in Interdisciplinary Mathematics, Volume 4.

Here is how "energy" itself is defined. Let Y be an oriented codimension one submanifold of Z. Let V be a vector field that is perpendicular to Y (in the Riemannian sense) and of unit length. Then

$$\int_Y T(V,V)(V \lrcorner dz) \tag{20.2}$$

is the *energy* in the electromagnetic field $d\theta$.

We can now show how this reduces to familiar formulas to be found in electromagnetic theory books. Let $Z = R^4$ = Minkowski space. Coordinates for Y are (t,x^i), $1 \leq i,j \leq m$. Let Z be the submanifold $\{t = 0\}$.

$$<dx^i, dx^j> = -\delta^{ij}$$

$$<dx^i, dt> = 0$$

$$<dt, dt> = 1 \quad .$$

(The velocity of light is normalized to be one.)

$$V = \frac{\partial}{\partial t}$$

$$F = E_1 \; dx^i \quad dt + B_{ij} dx^i \quad dx^j \tag{20.3}$$

$$<F,F> = E_i E_j <dx^i \quad dt, \; dx^j \quad dt> + B_{ij} B_k \; <dx^i \quad dx^j, \; dx^k \quad dx>$$

$$= - E_i E_j \delta^{ij} + B_{ij} B_{ij} - B_{ij} B_{ji}$$

$$= - E_i E^i + 2B_{ij} B^{ij}$$

$$= - \vec{E} \cdot \vec{E} + \vec{B} \cdot \vec{B} \quad ,$$

which is the traditional expression for the Lagrangian

$$\frac{\partial}{\partial t} \lrcorner \; F = -E_i \; dx^i$$

$$<\frac{\partial}{\partial t} \lrcorner \; F, \; \frac{\partial}{\partial t} \lrcorner \; F> = - E_i \; E^i$$

$$T\left(\frac{\partial}{\partial t}, \frac{\partial}{\partial t}\right) = \frac{1}{2} \; (\vec{E} \cdot \vec{E} + \vec{B} \cdot \vec{B}) \tag{20.4}$$

which is again, the usual Poynting formula for the energy of the electromagnetic field.

21. ZERO ENERGY ("GROUND STATE") ELECTROMAGNETIC FIELDS. "INSTANTONS"

Note that

$$T\left(\frac{\partial}{\partial t}, \frac{\partial}{\partial t}\right) \geq 0$$

(for *real* fields). Hence

$$(\text{total energy} = 0 \Rightarrow (\vec{E} = 0 = \vec{B} \text{ at } t = 0) \quad .$$

The Maxwell field equations (a hyperbolic system) now imply that

$$\vec{E} = 0 = \vec{B}$$

identically. However, if one now allows *complex* valued fields, there is another possibility:

$$\sqrt{-1}\, B = E \quad ,$$

or

$$2B_{ij} = \sqrt{-1}\, \varepsilon^{k}_{ij} E_k \quad . \tag{21.1}$$

Solutions of Maxwell's equations satisfying (21.1) are called *instantons*. It is readily seen that this means

$$*F = \sqrt{-1}\, F \tag{21.2}$$

where * is the Hodge dual operator for Minkowski space.

What is the geometric meaning of condition (21.1)? To see this, return to the *differential form* interpretation of the electromagnetic field

$$d\theta = E_i dx^i \wedge dt + B_{ij} dx^i \wedge dx^j$$

Now, consider x^i,t as *complex* variables. Set

$$t = \sqrt{-1}\ \tau \tag{21.3}$$

$$d \ = \sqrt{-1}\ E_i dx^i \wedge d\tau + B_{ij} dx^i \wedge dx^j$$

$$= E_i' dx^i \wedge d\tau + B_{ij}' dx^i \wedge dx^j$$

with

$$E_i' = \sqrt{-1}\, E_i \quad , \quad B_{ij}' = B_{ij} \quad .$$

This again has the "real" property, at the expense of involving the change of variable (21.3), which changes the Lorentz metric in R^4 to the *Euclidean* metric. Hence,

> Instantons may be considered as *real* self-dual solutions of the Maxwell equations in *Euclidean* R^4.

The point to this is that the Maxwell equations in Euclidean R^4 are an *elliptic* system of partial differential equations. Further, the Maxwell equations and the self-dual conditions are conformally invariant, hence the equations can be carried over to the *constant curvature* metric on the four-sphere, . which is a *compact* manifold. Now, the theory of elliptic partial differential equations on compact manifolds is the most extensively developed part of the discipline--powerful methods are available, culminating in the *Atiyah-Singer index theorem*. In fact, in this case--the Maxwell equations, i.e., Yang-Mills with an abelian structure group--the question can be settled with the prototypical theory of this sort, Hodge's theory of harmonic integrals. Since $H^2(S^4,R) = 0$, there are *no* nonzero solutions of the Euclidean Maxwell equations which are everywhere regular on the four-sphere.

Often, we find solutions of the equations in the Euclidean case which are *algebraic*. They can then be "analytically continued" over to Minkowski space. One then typically obtains solutions in terms of the space-time variables (x^μ) which are "algebraic" in these coordinates. It is interesting to note that a major topic in the 19th century theory of nonlinear partial differential equations was precisely this question of classifying the "algebraic" solutions.

22. THE YANG-MILLS GENERALIZATION

Everything done in the previous section for Maxwell can now be carried over to Yang-Mills. Let $\mathscr{K}$ be a real Lie algebra. Let

$$\mathscr{D}^r(Z,\mathscr{K})$$

be the r-th degree differential forms on Y *with values in* $\mathscr{K}$. They can be written in the form

$$\omega = \omega_1 \otimes A_1 + \dots + \omega_n \otimes A_n$$

$$\omega_1,\dots,\omega_n \in \mathscr{D}^r(Y) \quad ,$$

i.e., *scalar* differential forms

$$A_1,\dots,A_n \in \mathscr{K} .$$

Suppose, given a

$$\theta \in \mathscr{D}^1(Z,\mathscr{L}) \quad .$$

It determines a map

$$\nabla_\theta : \mathscr{D}^r(Z,\mathscr{L}) \to \mathscr{D}^{r+1}(Z,\mathscr{L})$$

called *covariant differentiation.* If

$$\theta = \theta_1 \otimes A_1 + \dots + \theta_n \otimes A_n \quad ,$$

$$\omega = \omega_1 \otimes B_1 + \dots + \omega_n \otimes B_n \quad ,$$

then

$$\nabla_\theta(\omega) = d\omega_1 \otimes B_1 + \dots + d\omega_n \otimes B_n \quad + \sum_{i,j=1}^{n} (\theta_i \quad \omega_j) \otimes A_i \quad B_j \tag{22.1}$$

Suppose also that < , > is the inner product on differential forms determined by a Riemannian metric on Y and that (,) is an inner product on $\mathscr{K}$ invariant under the adjoint representation (e.g., the Killing form). One can then define an inner product on $\mathscr{D}^r(Y,\mathscr{L})$ in the usual "tensor product" way:

$$\langle\theta,\omega\rangle = \sum \langle\theta_i,\omega_j\rangle(A_i,B_j) \quad . \tag{22.2}$$

Similarly, the Hodge dual operator * can be generalized:

$$dz\langle\theta,\omega\rangle = \theta \wedge (*\omega) \quad , \tag{22.3}$$

where

$$\theta \wedge \omega = \sum_{i,j} (\theta_i \wedge \omega_j)(A_i,B_j) \quad .$$

Alternately,

$$*\omega = \sum (*\omega_i) \times B_i \quad , \tag{22.4}$$

where $(*\omega_i)$ is the Hodge dual of ω_i as a scalar differential form (with respect to the Riemannian metric on Y) and B_i is an orthonormal basis of $\mathscr{K}$, i.e.,

$$(B_i,B_j) = \delta_{ij} \quad .$$

Exactly the same formulas now hold for Yang-Mills as for Maxwell. The field equations are

$$\nabla_\theta \theta = 0$$
$$\nabla_\theta (*\nabla_\theta \theta = 0 \qquad (22.5)$$

(The major difference is that now the "potential" cannot be eliminated from the equations!)

Again, if $Y = R^4 \equiv$ Minkowski space-time, we can write

$$\vec{E} = -\frac{\partial}{\partial t} \lrcorner \nabla_\theta \theta \quad , \quad \vec{B} = \nabla_\theta \theta - \vec{E} \ . \qquad (22.6)$$

These are the "electric" and "magnetic" components of the field. They are $\mathcal{K}$-valued, time-dependent differential forms on R^3.

$$D_\theta \theta = \vec{E} * dt + \vec{B} \quad , \quad *D_\theta \theta = (*\vec{E}) \pm (*\vec{B}) \wedge dt \ .$$

The *zero energy* solutions of the Yang-Mills field are those for which

$$\nabla_\theta \theta = 0 \ . \qquad (22.7)$$

The "instantons" are those which (possibly after complexification) are given as follows

$$*\nabla_\theta \theta = \lambda \nabla_\theta \theta \qquad (22.8)$$

with $\lambda \in \mathbb{C}$.

We have seen in the earlier sections how such forms may define connections in fiber bundles $\pi: X \to Z$ with Z as base space.

23. SOME LINEAR AND BILINEAR DIFFERENTIAL OPERATORS ON PRINCIPAL FIBER BUNDLES AND THEIR RELATION TO CONNECTIONS AND YANG-MILLS THEORY

Let

$$\pi: X \to Z$$

be a principal bundle with structure group K. Then, $\mathcal{K}$, the Lie algebra of K, acts freely as a Lie algebra of vector fields on X. In other words, $\mathcal{K}$ is a Lie subalgebra of X, and, for each nonzero $A \in \mathcal{K}$, $x \in X$,

$$A(x) \neq 0 \ .$$

We have seen that K-invariant connections for this bundle are determined by $\mathcal{K}$-valued differential forms on X, and the curvature of such a connection is determined by a *nonlinear* differential operator (reducing to a linear one, namely, exterior differentiation, when $\mathcal{K}$ is abelian) on such $\mathcal{K}$-valued forms. This operator is one of the basic operators in the Yang-Mills theory.

Now, this operator is nonlinear. However, its nonlinearity is "mild," in fact it is determined by a certain *bilinear* differential operator. In this section I want to develop a geometric "calculus" of such operators, based on certain ideas of *Vector Bundles in Mathematical Physics* and *Geometry, Physics and Systems*. My goal is also to lead into material on the theory of *characteristic* classes and its relation to *curvature*. One of the most interesting and useful features of casting physical theories in terms of *connections* is that the theory of characteristic classes [78] comes into play as a source of topological insight. If one wants a theory in which quarks are *automatically* confined, it is certainly plausible that a natural mechanism for this is that they are knotted together in some way.

Let us begin with a generalization of *Lie derivative*. Given

$$A \in \mathscr{D}^0(X,\mathscr{K}) \quad , \quad \omega \in \mathscr{D}^r(X,\mathscr{K}) \quad ,$$

$\mathscr{L}_A(\omega)$ is to be another element of $\mathscr{D}^r(X,\mathscr{K})$. $(A,\omega) \to \mathscr{L}_A(\omega)$ is to be a first order bilinear differential operator:

$$\mathscr{D}^0(X,\mathscr{K}) \times \mathscr{D}^r(X,\mathscr{K}) \to \mathscr{D}^r(X,\mathscr{K}) \quad .$$

Let us begin the definition of $\mathscr{L}$ with the case: $r = 0$.

Now, $\mathscr{D}^0(X,\mathscr{K})$ is the space of maps $X \to \mathscr{K}$. Geometrically, it is the space of *vertical vector fields* of the fiber space π: $X \to Z$, i.e., the vector fields V on X such that $\pi_*(V) = 0$. Now define:

$$\mathscr{L}_A(\omega) = [A,\omega] \quad , \tag{23.1}$$

where the bracket [,] on the right hand side of (23.1) is just the Jacobi bracket of the vector fields in VE(π) determined (using relation (23.1)) by V and ω.

Now, we can do the case of general r: For $V_1,\ldots,V_2 \in \mathscr{V}(X)$:

$$\mathscr{L}_A(\omega)(V_1,\ldots,V_r) = \mathscr{L}_A(\omega(V_1,\ldots,V_r)) - \omega([A,V_1],\ldots,V_r) - \ldots - \omega(V_1,\ldots,[A,V_r])$$

= , using the identification (23.1)

$$[A,\omega(V_1,\ldots,V_r)] - \omega([A,V_1],\ldots,V_r) - \ldots - \omega(V_1,\ldots,[A,V_r])$$

Now, let us look at the Maurer-Cartan operator introduced earlier, in order to define the curvature tensor. For $\theta \in \mathscr{D}^1(X,\mathscr{K})$, $V_1,V_2 \in \mathscr{V}(X)$,

$$D(\theta)(V_1,V_2) - V_1(\theta(V_2)) - V_2(\theta(V_1)) - \theta([V_1,V_2]) - [\theta(V_1),\theta(V_2)] \tag{23.2}$$

where:

$$[\theta(V_1),\theta(V_2)](x) = [\theta(V_1)(x),\theta(V_2(x)] \quad .$$

Note that this bracket is $\mathscr{F}(X)$-linear.

We can abstract from this a more general operator: For $\theta,\omega \in \mathscr{D}^1(X,\mathscr{K})$, $V_1V_2 \in \mathscr{V}(X)$,

$$\nabla_\theta\omega(V_1,V_2) = V_1(\omega(V_2)) - V_2(\omega(V_1) - \omega([V_1,V_2]) - \tfrac{1}{2}([\theta(V_1),\omega(V_2)] + [\theta(V_2),\omega(V_1)] \tag{23.3}$$

<u>Theorem 23.1</u>. The operator

$$(\theta,\omega) \to \nabla_\theta\omega$$

is a first order, bilinear differential operator

$$\mathscr{D}^1(X,\mathscr{K}) \times \mathscr{D}^1(X,\mathscr{K}) \to \mathscr{D}^2(X,\mathscr{K}) \quad .$$

The Maurer-Cartan operator (23.2) is given as follows in terms of it:

$$D\theta = \nabla_\theta\theta \quad .$$

As the notation indicates, ∇ is a variant of the *covariant derivative*.

The following algebraic rules are satisfied:

$$\nabla_{f\theta}\omega = f\nabla_\theta\omega \qquad \text{for } f \in \mathcal{F}(Z) \tag{23.4}$$

$$\nabla_\theta(f\omega) = df \quad \omega + f\nabla_\theta\omega \tag{23.5}$$

Let us write the operator in terms of a basis (A_a), $1 \leq a,b \leq m$, for $\mathcal{K}$, and scalar valued differential forms (θ^1,ω^1) such that

$$\theta = \theta^a A_a \quad , \quad \omega = \omega^a A_a \quad , \quad [A_a,A_b] = \lambda^c_{ab} A_c \quad .$$

<u>Theorem 23.2.</u>

$$\nabla_\theta\omega = (d\omega^a - \tfrac{1}{2}\lambda^a_{bc}\theta^b \quad \omega^c)A_a \quad . \tag{23.6}$$

<u>Proof.</u> Straightforward verification using (23.3).

Set:

$$\nabla^a_\theta\omega = d\omega^a - \tfrac{1}{2}\lambda^a_{bc}\,\theta^b \wedge \omega^c \quad . \tag{23.7}$$

Thus, (23.6) means that

$$\nabla_\theta\omega = \nabla^a_\theta\omega\, A_a \quad . \tag{23.8}$$

Also, set:

$$\Omega = \nabla_\theta\theta \tag{23.9}$$

We shall now derive differential relations satisfied by ∇_θ, by applying exterior differentiation. To avoid proliferating indices, it is useful to adopt a notation using vectors and matrices whose elements are differential forms

$$\tilde{\omega} = \begin{pmatrix} \omega^1 \\ \vdots \\ \omega^m \end{pmatrix}$$

$$\tilde{\theta} = \begin{pmatrix} \theta^1 \\ \vdots \\ \theta^m \end{pmatrix}$$

$$\tilde{\lambda}(\theta) = \tfrac{1}{2}\lambda^a_{bc}\theta^b)^m_{a,c=1}$$

$$d\tilde{\omega} = \begin{pmatrix} d\omega^1 \\ \vdots \\ d\omega^m \end{pmatrix}$$

Thus, we write (23.7) as:

$$\nabla_\theta\tilde{\omega} = d\tilde{\omega} - \tilde{\lambda}(\theta) \wedge \tilde{\omega} \tag{23.10}$$

(The multiplication $\wedge$ in the second term on the right hand side of (23.10) is a matrix multiplication, but with matrix elements in a non-commutative algebra. Note that $\tilde{\omega}$ is an $m\times 1$ matrix, $\tilde{\alpha}$ are $m\times m$.)

Now, we can very easily derive the differential relations that follow from (23.7) and (23.9). (They are the *Bianchi identities*.)

First, write (23.9) as:

$$\tilde{\Omega} = d\tilde{\theta} - \tilde{\lambda}(\theta) \wedge \tilde{\theta} \tag{23.11}$$

Then,

$$\begin{aligned} d\tilde{\Omega} &= dd\tilde{\theta} - d\tilde{\lambda}(\theta) \wedge \tilde{\theta} + \tilde{\lambda}(\theta) \wedge d\tilde{\theta} \\ &= 0 - d\tilde{\lambda}(\theta) \wedge \tilde{\theta} + \tilde{\lambda}(\theta) \wedge (\tilde{\Omega} + \tilde{\lambda}(\theta) \wedge \tilde{\theta}) \\ &= \text{, using (23.11)} \\ &\quad - (d\tilde{\lambda}(\theta) - \tilde{\lambda}(\theta) \wedge \tilde{\lambda}(\theta)) \wedge \tilde{\theta} + \tilde{\alpha} \wedge \tilde{\Omega} \end{aligned} \tag{23.12}$$

$d\tilde{\lambda}(\theta)$ can be expressed in terms of $\tilde{\lambda}(\theta)$ and $\tilde{\Omega}$. When these relations are substituted into (23.12), $d\tilde{\Omega}$ can be expressed in terms of $\tilde{\Omega}, \tilde{\theta}$, and the structure constants (λ^a_{bc}) of the Lie algebra $\mathcal{K}$. In order to keep working with matrix algebra (which is computationally convenient), set:

$$\tilde{\eta}(\Omega) = d\tilde{\lambda}(\theta) - \tilde{\lambda}(\theta) \wedge \tilde{\lambda}(\theta) \quad . \tag{23.13}$$

Thus, $\tilde{\eta}(\theta)$ is an m×m matrix of two-forms, which can be expressed (as the notation indicates) in terms of the Ω^a and the structure constants A^a_{bc}). It is *linear* in the Ω^a.

Combine (23.12) and (23.13)

$$d\tilde{\Omega} = \tilde{\lambda}(\theta) \wedge \tilde{\Omega} - \tilde{\eta}(\Omega) \wedge \tilde{\theta} \quad . \tag{23.14}$$

Let us think about this from the bilinear differential operator point of view. Define

$$\tilde{D}_1(\tilde{\theta}, \tilde{\omega}) = d\tilde{\omega} - \tilde{\lambda}(\theta) \wedge \tilde{\omega} \tag{23.15}$$

Then,

$$\tilde{\Omega} = \tilde{D}_1(\tilde{\theta}, \tilde{\theta}) \quad . \tag{23.16}$$

<u>Remark</u>. Keep in mind the physical significance of this. In the Maxwell case

$$\dim \mathcal{K} = 1 \quad ,$$

hence $\tilde{D}_1$ is a "degenerate" bilinear map, i.e., a "linear" one. ($\tilde{\theta}$ is the "potential," $\tilde{\Omega}$ the "electromagnetic field." In the Yang-Mills case, i.e., $\mathcal{K}$ non-abelian, we must carry along the *pair* $(\theta, \tilde{\Omega})$.

For the next step, forget that $\tilde{\Omega}$ is the curvature associated with θ. Think of it as any m×1 matrix of two-forms. Define:

$$\tilde{D}_3(\theta, \tilde{\Omega}) = \tilde{\lambda}(\theta) \wedge \tilde{\Omega} - \tilde{\eta}(\Omega) \wedge \tilde{\theta} \quad . \tag{23.17}$$

Thus, we have the following relation:

$$\tilde{D}_2(\theta, \tilde{D}_1(\theta, \theta)) = 0 \tag{23.18}$$

as another way of writing the Bianchi identities. Writing it this way emphasizes that it is a *nonlinear* generalization of the relation

$$dd = 0$$

that plays such a key role in differential geometry when one follows the methods of Cartan. The operator $\tilde{D}_2$ can also be thought of as *covariant differentiation* with respect to the connection θ.

Now let us be more general, and suppose only that X is a manifold and $\mathcal{K}$ is a Lie algebra over the real numbers as fields of scalars. (It is obviously interesting for physics to allow $\mathcal{K}$ to be infinite dimensional!) Let $\mathcal{D}^1(X, \mathcal{K})$, $\mathcal{D}^2(X, \mathcal{K})$, ...

be the differential forms on X with values in $\mathcal{K}$. We have defined a bilinear differential operator

$$D_1 : \mathcal{D}^1(X,\mathcal{K}) \times \mathcal{D}^1(X,\mathcal{K}) \to \mathcal{D}^2(X,\mathcal{K}) \qquad (23.19)$$

The following result holds.

<u>Theorem 23.2.</u> Let K be a connected Lie group whose Lie algebra is $\mathcal{K}$. (Thus we assume $\mathcal{K}$ is finite dimensional.) Let $\omega \in \mathcal{D}^1(K,\mathcal{K})$ be the left-invariant, K-invariant Cartan-Maurer form on K. Let $\theta \in \mathcal{D}^1(X,\mathcal{K})$. Then

$$D_1(\theta,\theta) = 0 \qquad (23.20)$$

if and only if the following condition is satisfied:

X can be covered by open subsets $\{U\}$, in each of which there is a map $\phi_U : X \to K$ such that:

$$\theta = \phi_U^*(\omega) \quad . \qquad (23.21)$$

Whenever $U, U' \in \{U\}$ intersect, there is a $k_{UU}' \in K$ such that

$$\phi_U = k_{UU'}\phi_{U'} \qquad (23.22)$$

The system $\{k_{UU'}\}$ defines a one-Cech cocycle with coefficients in K, hence a covering space X' of X, whose fiber is a coset space of the fundamental group $\pi_1(X)$ modulo a discrete subgroup which is isomorphic to a subgroup of K.

We now want to think of pairs

$$(\theta,\Omega) \in \mathcal{D}^1(X,\mathcal{K}) \times \mathcal{D}^2(X,\mathcal{K})$$

as a *Yang-Mills field* with K as structure group. The *first field equation* is:

$$D_1(\theta,\theta) = \Omega \quad . \qquad (23.23)$$

There is a bilinear operator

$$D_2 : \mathcal{D}^2)X,\mathcal{K}) \times \mathcal{D}^2(X,\mathcal{K}) \to \mathcal{D}^3(X,\mathcal{K}) \qquad (23.24)$$

called the *Bianchi operator*, such that

$$D_2(\theta,\Omega) = 0 \qquad (23.25)$$

whenever (θ,Ω) satisfy (23.23).

To complete a physical theory, we need another bilinear operator

$$\text{div}: \mathcal{D}^1(X,\mathcal{K}) \times \mathcal{D}^2(X,\mathcal{K}) \to \mathcal{D}^3(X,\mathcal{K}) \qquad (23.26)$$

Call this the *Hodge* or *generalized divergence operator*. The *Maxwell-Yang-Mills equations* are then:

$$D_1(\theta,\theta) = \Omega \qquad (23.27)$$

$$\text{div}(\theta,\Omega) = J \quad , \qquad (23.28)$$

where J is a given element of $\mathcal{D}^3(X,\mathcal{K})$. Of course, the "current" J may be given another field theory. Thus, the equations for this field, together with (23.27)-(23.28), will provide a *coupled, nonlinear field theory*. The Maxwell-Dirac equations of electrodynamics ("QED") are a more traditional prototype, with $\mathcal{K}$ abelian. The equations of modern elementary particle physics ("QCD") are a direct generalization, when K is a compact Lie group.

So far the space X has been left indeterminate. We have not even assumed that K acts on X. Let us then postulate that the following conditions are satisfied:

K acts freely on X

The orbit space K X = Z is a manifold

The quotient map $\pi: X \to K X = Z$ (which assigns to each $x \in X$ the orbit Kx on which it lies) is a submersion, i.e., (X,Z,π,K) is a *principal fiber bundle with structure group* K.

$\mathcal{K}$ is identified with the Lie algebra of vector fields on X, which results from letting the one parameter subgroup of K act on X.

Now, let

$$\theta: T(X) \to \mathcal{K} ,$$

i.e., $\theta \in \mathcal{D}^1(X,\mathcal{K})$ be a $\mathcal{K}$-valued one-form on X such that:

$$\theta(A(x)) = A \quad \text{for } A \in \mathcal{K} , \quad x \in X \tag{23.29}$$

The Lie derivative of θ by each element of $\mathcal{K}$ is zero. (23.30)

A θ satisfying (23.29)-(23.20) defines a K-*invariant connection* for the principal fiber bundle

$$(X,Z,\pi,K) \quad .$$

Let $\Omega \in \mathcal{D}^2(X,\mathcal{K})$ be its curvature, i.e.,

$$\Omega = D_1(\theta,\theta) \equiv \nabla_\theta \theta \tag{23.31}$$

Now, choose a bilinear differential operator

$$\text{div}: \mathcal{D}^1(\theta,\mathcal{K}) \times \mathcal{D}^2(\theta,\mathcal{K}) \to \mathcal{D}^3(\theta,\mathcal{K}) \tag{23.32}$$

A *Yang-Mills field* is a connection θ (for a fixed principal bundle structure) such that (23.31) is satisfied, plus the following condition:

$$\text{div}(\theta,\Omega) = J \quad ,$$

where J is a "given" element of $\mathcal{D}^3(X,\mathcal{K})$. (Of course, J could be determined by other equations.)

Let

$$*: \mathcal{D}^2(X,\mathcal{K}) \to \mathcal{D}^2(X,X)$$

be a *zero-th order differential operator*, not necessarily linear.

Remark. Thus, * is a fiber preserving mapping, not necessarily linear, of the following bundle over X

$$E = (T^d(C) \times T^d(X)) \otimes (X \times \mathcal{K})$$

into itself. The nonlinearity is required to cover generalizations of what the physicists call *nonlinear optics*.

Now, we can let

$$\text{div } D_3 = \nabla_\theta(*\Omega) \quad . \tag{23.35}$$

The Yang-Mills equations become:

$$\nabla_\theta \theta = \Omega \qquad \nabla_\theta(*\Omega) = J \tag{23.36}$$

BIBLIOGRAPHY

1. A. Barut, *Electrodynamics*, Macmillan, New York, 1964.
2. A. Belavin, A. Polyakov, A. Schwartz, and Y. Tyupkin, *Phys. Lett.* B 59, 85 (1975).
3. M. Born and H. Wolf, *Principles of Optics*, Bergeman Press, 1975.
4. R. Bott, "Topological Obstruction to Integrability," *Proc. Symp. Pure Math*, Vol. 16, 1975, Amer. Math. Soc., Providence, R.I.
5. E. Cartan, "Sur certaines expressions differentielles et le problème de Pfaff," *Ann. Ec. Norm.*, 16, 239-332 (1899).
6. E. Cartan, Les éspaces a connexion projective, *Conforme, Ann. Soc. Pol. Math.*, 2, 1923, 171-221.
7. E. Cartan, Les éspaces a connexion projective, *Bull. Soc. Math., France*, 52, 205-241, 1929.
8. E. Cartan, "Sur les varietés a connexion affine et la theorie de La Relativité Generalisée," *Ann. Ec. Norm.*, 4D, 32F 412, 1940.
9. E. Cartan, "Sue l'equivalence absolue de certains systemes d'equations differentielles," *Oevres Complètes*, Gauthier-Villars, Paris, 1953.
10. E. Cartan, "Sur l'integration de certains systèmes de pfaff de caractere deux," *Oevres Complètes*, Gauthier-Villars, Paris, 1953.
11. E. Cartan, *Les Systèmes Differentiels Exterieurs et leurs Applications Geometriques*, Herman, Paris, 1945.
12. E. Cartan, "Les systèmes de Pfaff a cinq variables," *Oevres Complètes, Pt. II*, Gauthier-Villars, Paris, 1953.
13. E. Corrigan and D.B. Fairlie, *Phys. Lett. B* 67, 69 (1977).
14. Y. Choquet-Bruhat, *Geometrie Differentielle et Systèmes Exterieurs*, Dunod, Paris, 1963.
15. J. Dieudonne, *Treatise on Analysis, Vol. 4*, Academic Press, New York, 1974.
16. C. Ehresmann, "Sur la topologie des certains éspaces homogenes," *Annals of Math* 35, 395-443 (1934).
17. C. Ehresmann, "Les connexions infinitesimales dans un éspace fibré," *Colloque de Topologie*, Bruxelles, 1950.
18. F. Estabrook and H. Wahlquist, "The geometric approach to sets of ordinary differential equations and Hamiltonian mechanics," *SIAM Rev.* 17, 201-220 (1975).
19. F. Estabrook and H. Wahlquist, "Prolongation structures of nonlinear evolution equations, I. *J. Math. Phys.* 16, 1-7 (1975).
20. F. Estabrook and H. Wahlquist, "Prolongation structures of nonlinear evolution equations," II, *J. Math. Phys.* 17, 1293-1297 (1976).
21. F. Estabrook and H. Wahlquist and R. Hermann, "Differential-geometric prolongations and Backlund transformations," in R. Hermann (ed.) *The Ames Research Center (NASA) 1976 Conference on the Geometric Theory of Non-Linear Waves*, Math Sci Press, Brookline, MA 1977.
22. F. Estabrook and H. Wahlquist, "Prolongation structures, connection theory and Backlund transformation," in F. Calogero (ed.), *Nonlinear Evolution Equations Solvable by the Spectral Transform, Research Notes in Mathematics, No. 26*, Pittman, London, 1978.
23. F. Estabrook, "Some old and new techniques for the practical use of differential forms," in R. Miura (ed.) *Backlund Transformation, the Inverse Scattering Method, Solitons and their Application, Lecture Notes in Mathematics, No. 515*, Springer-Verlag, Berlin, New York, 1976.
24. F. Estabrook and B.K. Harrison, "Geometric approach to inverse groups and solution of partial differential systems," *J. Math. Phys.* 12, 653-666 (1971).
25. R.B. Gardner, "Invariants of Pfaffian systems," *Trans. Am. Math. Soc.* 126, 514-543 (1967).
26. E. Goursat, *Lecons sur le Problème de Pfaff*, Herman, Paris, 1922.
27. V. Guillemin and S. Sternberg, *Geometric Asymptotics*, American Math. Soc., 1977.
28. R. Hermann, "Sur les isométries infinitésimaux et le groupe d'holonomie d'un espace de Riemann," *C.R. Acad. Sci., Paris*, 239, 1178-1180 (1954).
29. R. Hermann, "Sur les automorphismes infinitesimaux d'une G. structure, *C.R. Acad. Sci., Paris* 239, 1760-1761 (1954).
30. R. Hermann, "A sufficient condition that a map of Riemannian manifolds be a fiber bundle," *Proc. Am. Math. Soc.* 11, 236 (1960).

31. R. Hermann, "The differential geometry of foliations," *Ann. Math.* 72, 445-457, (1960).
32. R. Hermann, *Lie Groups for Physicists*, W.A. Benjamin, Reading, MA, 1966.
33. R. Hermann, "Equivalence of submanifolds of homogeneous spaces," *Math. Ann.* 158. 284-289 (1965).
34. R. Hermann, "Cartan connections and the equivalence problems for geometric structures," *Contributions to Differential Equations* 3, 199-248 (1964).
35. R. Hermann, "The differential geometry of foliations, II," *J. Math. Mech.* 11, 303-316 (1962).
36. R. Hermann, "Existence in the large of parallelism homomorphisms," *Trans. Am. Math. Soc.* 161, 170-183 (1963).
37. R. Hermann, "Formal linearization of Lie algebras of vector fields near an invariant submanifold," *J. Differential Geom.*, 1973.
38. R. Hermann, "A geometric formula for current algebra commutation relations," *Phys. Rev.* 177, 2449 (1969).
39. R. Hermann, "Quantum field theories with degenerate Lagrangians," *Phys. Rev.* 177, 2453 (1969).
40. R. Hermann, "Formal linearization of Lie algebras of vector fields near an invariant submanifold," *Trans. Am. Math. Soc.*, 1973.
41. R. Hermann, "Spectrum-generating algebras in classical mechanics, I and II," *J. Math. Phys.* 13, 833, 878 (1972).
42. R. Hermann, "Currents in classical field theories," *J. Math. Phys.* 13, 97 (1972).
43. R. Hermann, "Current algebra, Sugawara model and differential geometry," *J. Math. Phys.* 11, 1825-1829 (1970).
44. R. Hermann, *Fourier Analysis on Groups and Partial Wave Analysis*, W.A. Benjamin, New York, 1969.
45. R. Hermann, "Infinite dimensional Lie algebra and current algebra," *Proc. 1969 Battele-Seattle Rencontres on Math. Phys.*, Springer-Verlag, Berlin, 1970.
46. R. Hermann, *Lie Algebras and Quantum Mechanics*, W.A. Benjamin, New York, 1970.
47. R. Hermann, *Vector Bundles in Mathematical Physics, I and II*, W.A. Benjamin, New York, 1970.
48. R. Hermann, *Geometry, Physics and Systems*, M. Dekker, New York, 1973.
49. R. Hermann, *Gauge Fields and Cartan-Ehresmann Connections*, Part A, *Interdisciplinary Mathematics*, Vol. 10, Math. Sci. Press, Brookline, MA, 1975.
50. R. Hermann, "Modern differential geometry in elementary particle physics," *VII GIFT Conf. on Theoretical Physics, Salamanca, Spain, 1977*, A. Azcarraga (ed.), Springer-Verlag, Berlin.
51. R. Hermann, *Quantum and Fermión Differential Geometry*, Part A, Math Sci Press, Brookline, MA, 1977.
52. R. Hermann, *Geometry of Non-Linear Differential Equations, Backlünd Transformations and Solitons*, Part A, *Interdisciplinary Mathematics*, Vol. 12. Math. Sci. Press, Brookline, MA, 1976.
53. R. Hermann (ed.), *Sophus Lie's 1880 Transformation Group Paper*, comments and additional material by R. Hermann, Math Sci Press, Brookline, MA, 1975.
54. R. Hermann (ed.), *Ricci and Levi-Civita's Tensor Analysis Paper*, translation, comments, and additional material by R. Hermann, Math Sci Press, 1975.
55. R. Hermann (ed.), *Sophus Lie's 1884 Differential Invariants Paper*, comments and additional material by R. Hermann, Math Sci Press, 1975.
56. R. Hermann (ed.), *Proc. Ames (NASA) 1976 Conf. on the Geometry of Solitons*, Math Sci Press, Brookline, MA, 1977.
57. R. Hermann, "The Lie-Cartan geometric theory of differential equations and scattering theory," *Proc. 1977 Park City, Utah, Conf. on Differential Equations*, C. Byrnes (ed.).
58. R. Hermann, "Backlund transformations, and Lie theory as algorithms for solving and understanding nonlinear differential equations," *in Solitons in Action*, K. Lonngren and A. Scott (eds.), Academic Press, New York, 1978.
59. R. Hermann, *Toda Lattices, Cosymplectic Manifolds, Backlünd Transformations and Kinks*, Parts A and B, Math Sci Press, Brookline, MA, 1977.
60. R. Hermann, *Quantum and Fermion Differential Geometry*, Part A, Math Sci Press, Brookline, MA, 1977.
61. R. Hermann, *Fourier Analysis on Groups and Partial Wave Analysis*, W.A. Benjamin, Reading, MA, 1969.

62. R. Hermann, *Energy-Momentum Tensors*, Interdisciplinary Mathematics, Vol. 4, Math Sci Press, Brookline, MA, 1973.
63. R. Hermann, *Topics in General Relativity*, Interdisciplinary Mathematics, Vol. 5, Math Sci Press, Brookline, MA, 1973.
64. R. Hermann, *Topics in the Mathematics of Quantum Mechanics*, Interdisciplinary Mathematics, Vol. 6, Math Sci Press, Brookline, MA 1973.
65. R. Hermann, *Lectures on Mathematical Physics, II*, W.A. Benjamin, Reading, MA, 1972.
66. R. Hermann, *Yang-Mills, Naluza-Klein, and the Einstein Program*, Interdisciplinary Mathematics, Vol. 19, Math Sci Press, Brookline, MA, 1978.
67. R. Hermann, *Differential Geometry and the Calculus of Variations*, 2nd Ed., Interdisciplinary Mathematics, Vol. 17, Math Sci Press, Brookline, MA, 1977.
68. R. Hermann, "E. Cartan's geometric theory of partial differential equations," *Advances in Math.* 1, 265-317 (1965).
69. W.V. Hodge, *Theory and Application of Harmonic Integrals*, Cambridge Univ. Press, 1941.
70. G. Hooft, *Phys. Rev. D* 14, 3432 (1976).
71. J.D. Jackson, *Classical Electrodynamics*, Wiley, 1962.
72. M. Jimbo, T. Miwa, K. Vend, "Monodromy preserving deformation of linear ordinary differential equations with rational coefficients," *Physica* 2D 306-352 (1981).
73. K. Kodaira and D.C. Spencer, "On deformation of complex analytic structures," *Am. M. Math.* 67, 328-446 (1958).
74. S. Kobayaski, *Transformation Groups in Differential Geometry*, Springer-Verlag, Berlin, 1972.
75. S. Kobayaski and K. Nomizu, *Foundation of Differential Geometry*, Vols. I and II, Interscience, New York, 1963.
76. K. Kondo, *The RAAG Memoirs*, Tokyo, 1955.
77. G. Kron, "Non-Riemannian geometry of rotating electrical machinery," *J. Math. Mech.* 13, 103-194 (1934).
78. J. Milnor and S. Stasheff, *Characteristic Classes*, Princeton Univ. Press, 1965.
79. B. O'Neill, "Submersions and geodesics," *Duke Math. J.* 34, 363-373 (1967).
80. B. Reinhart, "Foliated manifolds with bundle-like metrics," *Ann. of Math.* 69, 119-132 (1959).
81. G. Reeb, *Sur Certaines Proprietes Topologiques des Varieté Feuilletees*, Herman, Paris, 1952.
82. G. de Rham, *Varietés Differentiables*, Herman, Paris, 1960.
83. W. Slebodzinski, *Exterior Forms and their Applications*, Polish Scientific Publishers, Warsaw, 1970.
84. M. Spivak, *Calculus on Manifolds*, W.A. Benjamin, New York, 1965.
85. M. Spivak, *A Comprehensive Introduction to Differential Geometry* Brandeis Univ., Waltham, MA, 1970.
86. N. Steenrod, *The Topology of Fiber Bundles*, Princeton Univ. Press, 1951.
87. S. Sternberg, *Lectures on Differential Geometry* Prentice-Hall, Englewood Cliffs, NJ, 1964.
88. A. Trautman, "Fiber Bundles, Gauge Fields, and Gravitation," *General Relativity and Gravitation*, Vol. I, A. Held, (Ed.), Plenum Pub. Co., 1980.
89. N. Wallach, *Symplectic Geometry and Fourier Analysis*, Math Sci Press, 1977.

A Geometric Introduction to Yang-Mills-Equations

by

Thomas Friedrich
Sektion Mathematik
Humboldt-Universität zu Berlin
DDR - 1086 Berlin, PSF1297

1. The mathematical formulation of the Yang-Mills-Equation.

Let M^n be a compact oriented Riemannian manifold, G a compact Lie group and denote by $\mathfrak{g}$ the Lie algebra of G. Consider a principal G-fibre bundle $(P,\pi,M^n;G)$ over M^n. A gauge transformation is a fibre-preserving diffeomorphism $f:P \longrightarrow P$ such that $f(g\cdot p) = g\cdot f(p)$. The group of all gauge transformations we denote by $\mathcal{G}(P)$. A connection (gauge potential) is a 1-form $\alpha : TP \longrightarrow \mathfrak{g}$ on P with values in $\mathfrak{g}$ such that

1.) $L_g^* \alpha = Ad(g)\alpha$

2.) If $X \in \mathfrak{g}$ and if $\tilde{X}$ is the corresponding vertical vector field on P, then $\alpha(\tilde{X}) = X$.

The set $\mathcal{C}(P)$ of all connections in P is an affine space. In fact, the difference $\alpha - \alpha'$ of two connections α, α' has the following properties: a.) $L_g^*(\alpha - \alpha') = Ad(g)(\alpha - \alpha')$

b.) $\alpha - \alpha' = 0$ on vertical vectors.

Hence $\alpha - \alpha'$ is a 1-form on M^n with values in the vector bundle $\underline{\mathfrak{g}} = P\times_{Ad}\mathfrak{g}$, $\alpha - \alpha' \in \Gamma(T^* \otimes \underline{\mathfrak{g}})$. The group $\mathcal{G}(P)$ acts on $\mathcal{C}(P)$ by $f\cdot\alpha = (f^{-1})^*\alpha$ and we denote by $\mathcal{M}(P) = \mathcal{C}(P)/\mathcal{G}(P)$ the corresponding orbit space.

The curvature form $\Omega = D\alpha$ of a connection α is an Ad-equivariant 2-form on P vanishing on vertical vectors. Hence we can understand Ω as a 2-form on M^n with values in $\underline{\mathfrak{g}}$, $\Omega \in \Gamma(\Lambda^2 \otimes \underline{\mathfrak{g}})$.

Now we fix an Ad-invariant inner product $(\ ,\ )$ in the Lie algebra $\mathfrak{g}$ and we define the length of Ω by

$$|\Omega|_m^2 = \sum_{i<j} |\Omega(e_i,e_j)|^2$$

where $e_1,\dots,e_n$ is an orthonormal base at $m \in M^n$.

This lecture was given at the fifth Scheveningen Conference on Differential Equations, August 23-28 1981.

On the space $\mathcal{C}(P)$ of connections we consider the Yang-Mills functional L: $\mathcal{C}(P) \longrightarrow [0, \infty)$

$$L(\alpha) = \frac{1}{2} \int_{M^n} |\Omega|^2 dM^n .$$

If the connections α and α' are equivalent, then $L(\alpha) = L(\alpha')$ holds. This means that L is a functional on the orbit space $\mathcal{M}(P)$, $L : \mathcal{M}(P) \longrightarrow [0, \infty)$.

Definition: A critical point α of the Yang-Mills functional L is called a Yang-Mills field.

To obtain the differential equation defining Yang-Mills fields we need some differential operators. A connection α introduces a covariant derivation $\nabla^\alpha : \Gamma(\mathfrak{g}) \longrightarrow \Gamma(T^* \otimes \mathfrak{g})$ in the associated vector bundle $\mathfrak{g} = P \times_{Ad} \mathfrak{g}$. If $\psi^p = \varphi^p \otimes s \in \Gamma(\Lambda^p \otimes \mathfrak{g})$ is a p-form with values in $\mathfrak{g}$, we define

$$d^\alpha \psi^p = d\varphi^p \otimes s + (-1)^p \varphi^p \otimes \nabla^\alpha s .$$

Then $d^\alpha : \Gamma(\Lambda^p \otimes \mathfrak{g}) \longrightarrow \Gamma(\Lambda^{p+1} \otimes \mathfrak{g})$ is a first order differential operator and we denote by δ^α the L^2- adjoint operator. Using d^α the Bianchi identity can be written in the form $d^\alpha \Omega = 0$.

Proposition: α is a Yang-Mills field if and only if $\delta^\alpha \Omega = 0$.

In fact, if α_t is a family of connections with $\alpha_0 = \alpha$ and $A_t = \alpha_t - \alpha \in \Gamma(T^* \otimes \mathfrak{g})$, $B = \frac{d}{dt}(A_t)_{t=0} \in \Gamma(T^* \otimes \mathfrak{g})$, then

$\Omega_t = \Omega + d^\alpha A_t + \frac{1}{2}[A_t \wedge A_t]$. Hence we obtain

$$\frac{d}{dt} L(\alpha_t)_{t=0} = \frac{1}{2} \int_{M^n} \frac{d}{dt} |\Omega_t|^2 dM = \int_{M^n} \left\langle \frac{d\Omega_t}{dt} , \Omega_t \right\rangle_{t=0} =$$

$$= \int_{M^n} \langle d^\alpha B , \Omega \rangle dM = \int_{M^n} \langle B , \delta^\alpha \Omega \rangle dM .$$

This proves that α is a critical point of L iff $\delta^\alpha \Omega = 0$.

Remark: Denote by $* : \Lambda^p \longrightarrow \Lambda^{n-p}$ the Hodge operator. Then $\delta^\alpha = (-1)^{p+1} * d^\alpha *$ and the Yang-Mills equation $\delta^\alpha \Omega = 0$ is equivalent to the following equation: $d^\alpha(*\Omega) = 0$. A connection α on a 4-dimensional manifold M^4 is called self-dual, if $*\Omega = \Omega$. Using the Bianchi identity we have $d^\alpha(*\Omega) = d^\alpha(\Omega) = 0$ in case α is self-dual. Hence every self-dual connection on a 4-dimensional manifold M^4 is a Yang-Mills field.

Now we state a formula for the second variation of the Yang-Mills functional at a critical point. To do this we define first of all a bundle morphism $k^\alpha : T^* \otimes \mathfrak{g} \longrightarrow T^* \otimes \mathfrak{g}$ by

$$(k^{\alpha}\varphi)(X) = \sum_{i=1}^{n} \left[\Omega(e_i,X)\ ,\ \varphi(e_i)\right]$$

where $e_1, \dots, e_n$ is an orthonormal base and $[\ ,\ ]$ denotes the bracket in the Lie algebra $\mathfrak{g}$.

Proposition (see [3]): Let $\alpha \in \mathcal{C}(P)$ be a Yang-Mills field ($\delta^{\alpha}\Omega = 0$) and α_t a variation of α . Denote the variation field by $B = \frac{d}{dt}(\alpha_t - \alpha)_{t=0} \in \Gamma(T^* \otimes \mathfrak{g})$. Then

$$\frac{d^2}{dt^2} L(\alpha_t)_{t=0} = \int_{M^n} \langle \delta^{\alpha} d^{\alpha} B + k^{\alpha}(B)\ ,\ B \rangle \quad .$$

Finally we consider the case of dimension four. There the Hodge operator $*: \Lambda^2 \longrightarrow \Lambda^2$ is an involution ($** = 1$) and we get a decomposition $\Lambda^2 = \Lambda^2_+ + \Lambda^2_-$ into the (± 1) - eigenspaces of $*$. The curvature form Ω decomposes into $\Omega = \Omega_+ + \Omega_-$ and we have

$$L(\alpha) = \frac{1}{2} \int_M |\Omega|^2 \geq \frac{1}{2} \int_M |\Omega_+|^2 - |\Omega_-|^2$$

$$L(\alpha) = \frac{1}{2} \int_M |\Omega|^2 \geq \frac{1}{2} \int_M - |\Omega_+|^2 + |\Omega_-|^2 .$$

On the other hand, the Pontrjagin number of the vector bundle $\mathfrak{g}$ is given by

$$p_1(\mathfrak{g}) = - c_2(\mathfrak{g} \otimes C) = \frac{1}{4\pi^2} \int_M |\Omega_+|^2 - |\Omega_-|^2$$

and we obtain the inequality $L(\alpha) \geq 2\pi^2 |p_1(\mathfrak{g})|$. Hence in case of dimension four we have a topological lower bound for the Yang-Mills functional and equality $L(\alpha) = 2\pi^2 |p_1(\mathfrak{g})|$ appears if and only if α is self-dual ($\Omega_- = 0$) or anti-self-dual ($\Omega_+ = 0$).

2. The U(1) - theory and a theorem of A. Weil.

In this section we study the case of the abelian group $G = U(1) = S^1$. Let $\pi : P \longrightarrow M^n$ be a principal S^1-fibre bundle. Then the gauge group $\mathcal{G}(P)$ is isomorphic to the group of all maps $\bar{f} : M^n \longrightarrow S^1$ since every gauge transformation $f: P \longrightarrow P$ is given by $f(p) = \bar{f}(\pi(p))\cdot p$. The Lie algebra of S^1 is $i\cdot R$ and the curvature form Ω of a connection α is an invariant form on P : $L_g^* \Omega = Ad(g)\Omega = \Omega$. Hence Ω is a usual 2-form on M^n with values in $i\cdot R$. Instead of Ω we consider $\bar{\Omega} = \Omega / 2\pi i$. Then $\bar{\Omega}$ is a reell, closed 2-form on M^n which represents the first Chern class $c_1^R(P) \in H^2(M^n;R)$ of P in the cohomology with reell coefficients: $c_1^R(P) = [\bar{\Omega}]$. Now we define the set

$$\mathcal{F}(P) = \left\{ w : w \text{ is a 2-form such that } dw=0 \text{ and } [w] = c_1^R(P) \right\}.$$

Then we obtain a map $\mathcal{C}(P) \longrightarrow \mathcal{F}(P)$ given by $\alpha \longrightarrow \overline{\Omega}$. Since in case of $G = U(1)$ two gauge equivalent connections have the same curvature, this map induces a map $\mathcal{M}(P) \longrightarrow \mathcal{F}(P)$.

Theorem of Weil (see for example [9]): $\mathcal{M}(P) \longrightarrow \mathcal{F}(P)$ is surjective and a principal $Pic(M^n)$ - fibre bundle, where $Pic(M^n) = H^1(M,R)/H^1(M,Z)$.

Now we consider the commutative diagram

$$\begin{array}{ccccc} \mathcal{C}(P) & \longrightarrow & \mathcal{M}(P) & \longrightarrow & \mathcal{F}(P) \\ & \searrow^{L} & \downarrow L & \swarrow^{\overline{L}} & \\ & & R^1 & & \end{array}$$

where $\overline{L} : \mathcal{F}(P) \longrightarrow R^1$ is defined by $L(w) = 2\pi^2 \int_M |w|^2$. Applying the Hodge theory we see that $\overline{L}$ has only one critical point, namly the unique harmonic form $w_{harm} \in \mathcal{F}(P)$ and this critical point is an absolute minimum for $\overline{L}$. This implies

Corollary: The set of all critical points of the Yang-Mills functional $L : \mathcal{M}(P) \longrightarrow R^1$ is diffeomorphic to $Pic(M^n)$ and all critical points are absolute minima. Furthermore

$$\inf_{\alpha \in \mathcal{M}(P)} L(\alpha) = 2\pi^2 \int_M |w_{harm}|^2$$

where w_{harm} is the unidue harmonic form satisfying $[w_{harm}] = c_1^R(P)$.

3. Some stability theorems.

In contrast to the abelian case the question about the structure of the set of all critical points in the general case is a difficult one. J.P.Bourguignon and H.B.Lawson [3], [4] proved that in case of a small group G (= SU(2) , U(2) , SO(3) , ..) there are no other local minima than absolute minima. In this section we shall formulate one result in this direction.

Definition: A Yang-Mills field α ($\delta^\alpha \Omega = 0$) is called weakly stable , if for every variation α_t

$$\frac{d^2}{dt^2} L(\alpha_t)_{t=0} \geq 0 \qquad \text{is valid.}$$

Theorem ([3], [4]):

Let $M^4 = K/H$ be a compact, oriented homogeneous Riemannian manifold and suppose that the quadratic form $H^2(M^4;R) \times H^2(M^4;R) \longrightarrow R$ is positive definite (for example S^4, $P^2(C)$ or $S^1 \times (S^3/\Gamma)$ where S^3/Γ is a homogeneous space of constant curvature). Then every weakly stable Yang-Mills field with G = SU(2) is (anti-)self-dual and hence an absolute minimum.

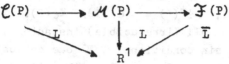

4. The moduli-space of instantons.

Let G be a simple compact Lie group. An instanton is a self-dual ($*\Omega = \Omega$) connection in a principal G-fibre bundle over a 4-dimensional manifold M^4. Every instanton α is a Yang-Mills field and realizes the absolute minimum of the Yang-Mills functional, $L(\alpha) = 2\pi^2/p_1(\mathfrak{g})/$. Denote by $\mathcal{C}^+(P) \subset \mathcal{C}(P)$ the set of all irreducible instantons in the bundle P and let $\mathcal{M}^+(P) \subset \mathcal{M}(P)$ be the corresponding orbit space, $\mathcal{M}^+(P) = \mathcal{C}^+(P)/\mathcal{G}(P)$. $\mathcal{M}^+(P)$ is called the moduli space of all (irreducible) instantons. $\mathcal{M}^+(P)$ is a manifold and under certain conditions at M^4 one can calculate the dimension of this moduli space. To formulate the main result in this direction, we need some notations.

consider an oriented 4-dimensional Riemannian manifold M^4 and an orthonormal reper. There we define the Weyl-tensor W by the formula

$$W_{ijkl} = R_{ijkl} + \frac{1}{2}(R_{ik}\delta_{jl} + R_{jl}\delta_{ik} - R_{il}\delta_{jk} - R_{jk}\delta_{il}) + \\ + \frac{\tau}{6}(\delta_{jk}\delta_{il} - \delta_{jl}\delta_{ik})$$

where R_{ij} are the components of the Ricci-tensor and τ is the scalar curvature. Due to the rule

$$W(e_i \wedge e_j) = \frac{1}{2}\sum_{k,l} W_{ijkl}e_k \wedge e_l$$

we understand the Weyl-tensor as a morphism $W : \Lambda^2 \longrightarrow \Lambda^2$. With respect to the decomposition $\Lambda^2 = \Lambda^2_+ + \Lambda^2_-$ this tensor has the following simple block form

$$W = \begin{pmatrix} W_+ & 0 \\ 0 & W_- \end{pmatrix} \quad , \quad W_\pm : \Lambda^2_\pm \longrightarrow \Lambda^2_\pm \quad .$$

The Riemannian manifold M^4 is called self-dual if $W_- = 0$.

<u>Remark:</u> The Levi-Civita connection of M^4 is self-dual ($*\Omega = \Omega$) if and only if $W_- = 0$ <u>and</u> $\tau = 0$.

S^4, $P^2(C)$, $S^1 \times S^3/\Gamma$, $S^2(k) \times H^2(-k)$ are examples of self-dual Riemannian manifolds.

<u>Theorem (see [1]):</u> Let M^4 be a closed self-dual Riemannian manifold with non-negative scalar curvature, $\tau \geq 0$, $\tau \neq 0$. Then $\mathcal{M}^+(P)$ is either empty or a manifold of dimension

$$\dim \mathcal{M}^+(P) = 2\cdot p_1(\mathfrak{g}) - \frac{1}{2}\dim(G)\cdot(\chi(M^4) - \sigma(M^4)).$$

In this formula $\chi(M^4)$ denotes the Euler characteristic and $\sigma(M^4)$ the signature of M^4.

Let us explain the calculation of $\dim \mathcal{M}^+(P)$. Suppose $\mathcal{M}^+(P) \neq$ and consider a family α_t of self-dual connections. Then

$$\Omega_t = \Omega + d^{\alpha} A_t + \frac{1}{2}[A_t \wedge A_t] \quad .$$

Since every connection α_t is self-dual ,

$$p_-(d^{\alpha} A_t + \frac{1}{2}[A_t \wedge A_t]) = 0$$

$$p_-(d^{\alpha} \frac{dA_t}{dt}_{t=0}) = 0$$

follows, where $p_- : \Lambda^2 \longrightarrow \Lambda^2_-$ is the projection corresponding to the decomposition. This means that the variation field $B = d\ A_t/dt$ defines an element in the first cohomology group of the elliptic complex

$$\Gamma(\mathfrak{g}) \xrightarrow{d^{\alpha}} \Gamma(\Lambda^1 \otimes \mathfrak{g}) \xrightarrow{p_- d^{\alpha}} (\Gamma^2_- \otimes \mathfrak{g}) \ ,$$

$[B] \in H^1(\mathfrak{g})(\alpha)$. Furthermore, it is not hard to check that if α_t are gauge equivalent to α ($\alpha_t = f_t \cdot \alpha$, $f_t \in \mathcal{G}(P)$) , then $[B] = 0$. This observation yields that the tangent space $T_{[\alpha]}\mathcal{M}^+(P)$ of $\mathcal{M}^+(P)$ at $[\alpha] \in \mathcal{M}^+(P)$ is isomorphic to $\ker(p_- d^{\alpha})/\mathrm{Im}(d^{\alpha}) = H^1(\mathfrak{g})(\alpha)$. On the other hand, since α is an irreducible connection and G is semisimple, $H^0(\mathfrak{g})(\alpha) = 0$ results. Using an integral formula of the Weitzenböck type one proves that $W_- = 0$ and $\tau \geq 0$, $\tau \neq 0$ implies $H^2(\mathfrak{g})(\alpha) = 0$. Now we calculate the index of the elliptic complex:

$$\dim \mathcal{M}^+(P) = \dim H^1(\mathfrak{g})(\alpha) = -\ (\text{a-index})(\mathfrak{g}) = -\ (\text{t-index})(\mathfrak{g}) =$$
$$= 2\cdot p_1(\mathfrak{g}) - \frac{1}{2}\dim(G)\cdot(\chi(M^4) - \sigma(M^4)).$$

<u>Remark:</u> Since $T_{[\alpha]}\mathcal{M}^+(P) = H^1(\mathfrak{g})(\alpha)$ the manifold $\mathcal{M}^+(P)$ has , in a natural manner, a Riemannian structure. Consider the $\mathcal{G}(P)$ - - bundle $\mathcal{C}^+(P) \longrightarrow \mathcal{M}^+(P)$. Then the choice $\alpha \longrightarrow H^1(\mathfrak{g})(\alpha)$ is a connection in this $\mathcal{G}(P)$ - bundle and $\mathcal{C}^+(P) \longrightarrow \mathcal{M}^+(P)$ becomes a Riemannian submersion. Whats about the geometry of $\mathcal{M}^+(P)$?

Let us now consider the case $M^4 = S^4$. Since $\pi_3(G) = Z$ all G - bundles over S^4 are classified by integers. We denote by P_k the G - bundle corresponding to $k \in Z = \pi_3(G)$. Then the following table contains the information whether $\mathcal{M}^+(P)$ is not empty and contains the dimension of the moduli space (see [1]):

G	$\mathcal{M}^+(P_k) \neq \emptyset$ iff	$\dim \mathcal{M}^+(P_k)$
SU(n)	$k \geq n$	$4nk - (n^2-1)$
Spin(n)	$k \geq \frac{1}{2}\cdot n$	$4(n-2)k - \frac{1}{2}n(n-1)$
Sp(n)	$k \geq \frac{1}{4}\cdot n$	$4(n+1)k - n(2n+1)$

5. SU(2) - Instantons on S^4.

Let $f: R^4 \longrightarrow S^4$, $f(x) = (2x , /x/^2 - 1)/(1+/x/^2)$, be the inverse map of the stereographic projection. Then

$$f^*(dS^4) = \left(\frac{2}{1+/x/^2}\right)^4 dR^4$$

holds. If $P \longrightarrow S^4$ is a G - bundle over R^4 with connection α , then $\bar{P} = f^*(P)$ is a G - bundle over R^4 with connection $\bar{\alpha} = f^*(\alpha)$. The length of the curvature forms are related by

$$/\Omega/^2_{S^4} = \left(\frac{1+/x/^2}{2}\right)^4 /\bar{\Omega}/^2_{R^4}$$

and hence we obtain

$$\int_{S^4} /\Omega/^2 dS^4 = \int_{R^4} /\bar{\Omega}/^2 dR^4 .$$

On the other hand, if $(\bar{P} , \bar{\alpha})$ is a G - bundle with a Yang - Mills field $\bar{\alpha}$ and if the curvature form is square integrable over R^4 , then there exists (P , α) on S^4 such that $f^*(P , \alpha) = (\bar{P} , \bar{\alpha})$ (see [12]). Hence Yang - Mills fields on S^4 correspond to Yang - -Mills fields on R^4 with square integrable curvature form.

Denote by H the algebra of quaternions and let Sp(1) be the group $Sp(1) = \{x \in H : /x/ = 1\}$. The group Sp(1) is isomorphic to SU(2) and the Lie algebra <u>sp(1)</u> of Sp(1) consists of all quaternions x such that Re(x) = 0 , <u>su(2)</u> = <u>sp(1)</u> = $Im(H) = \{x \in H : Re(x) = 0\}$. In <u>sp(1)</u> we choose the inner product by the condition that i,j,k are orthogonal and $/i/^2 = /j/^2 = /k/^2 = 2$. Now consider the connection $\bar{\alpha}_o : TR^4 = TH \longrightarrow$ <u>sp(1)</u>

$$\bar{\alpha}_o(x) = Im\left(\frac{x \cdot d\bar{x}}{1+/x/^2}\right) .$$

Then we calculate the curvature of this connection and obtain

$$\bar{\Omega}_o = - \frac{2}{(1+/x/^2)^2} \Big\{ (dx^1 \wedge dx^2 + dx^3 \wedge dx^4) \cdot i + (dx^1 \wedge dx^3 + dx^4 \wedge dx^2) \cdot j + (dx^1 \wedge dx^4 + dx^2 \wedge dx^3) \cdot k \Big\} .$$

From this formula it follows that $*\bar{\Omega}_o = \bar{\Omega}_o$ and $/\bar{\Omega}_o/^2 = \frac{48}{(1+/x/^2)^4}$. This means that $\bar{\alpha}_o$ is a (self-dual) Yang-Mills filed on R^4 with square integrable curvature.Hence $\bar{\alpha}_o$ is the restriction of a self-dual connection α_o in a SU(2)-bundle $P_k \longrightarrow S^4$. Let us calculate the Pontrjagin number k. Consider the vector bundle

$E_k = P_k \times_{SU(2)} C^2$. Then $p_1(E_k) = (c_1^2 - 2c_2)(E_k) = -2c_2(E_k) = 2k$.
On the other hand, since α_o is self-dual we get

$$4\pi^2 p_1(E_k) = \int_{S^4} |\Omega_+|^2 - |\Omega_-|^2 = \int_{S^4} |\Omega_o|^2 = \int_{R^4} |\bar{\Omega}_o|^2 =$$

$$= \int_{R^4} \frac{48}{(1+|x|^2)^4}\, dR^4 = 8\pi^2$$

and $k = 1$ follows. This means that $\bar{\alpha}_o$ defines a self-dual Yang-Mills field α_o in the SU(2) - bundle $P_1 \longrightarrow S^4$.
The connected component $\mathrm{Conf}_o(S^4)$ of the conformal group of S^4 is isomorphic to $\mathrm{Conf}_o(S^4) = SL(2,H)/\{\pm 1\}$. Here SL(2,H) acts on $S^4 = H \cup \{\infty\}$ by the rule

$$\begin{pmatrix} a & b \\ c & d \end{pmatrix} x = (ax + b)(cx + d)^{-1} \quad .$$

The action of $\mathrm{Conf}_o(S^4)$ on S^4 lifts to an action of this group on P_1. Since the Hodge operator $*$ is conformally invariant, the group $\mathrm{Conf}_o(S^4)$ acts on the moduli-space $\mathcal{M}^+(P_1)$,
<u>Theorem (see [1]):</u> $\mathrm{Conf}_o(S^4)$ acts transitively on $\mathcal{M}^+(P_1)$ and the isotropic group of $\alpha_o \in \mathcal{M}^+(P_1)$ is $SO(5) = Sp(2)/\{\pm 1\}$. Hence $\mathcal{M}^+(P_1) = SL(2,H)/Sp(2)$.
To obtain an explicit formula for an arbitrary SU(2)-instanton in the bundle P_1 we parametrize the space SL(2,H)/Sp(2) by $R_+^1 \times H \ni (\lambda, b)$

$$\longrightarrow \begin{pmatrix} 0 & \sqrt{\lambda} \\ -\frac{1}{\sqrt{\lambda}} & \frac{b}{\sqrt{\lambda}} \end{pmatrix} \cdot Sp(2) \in SL(2,H)/Sp(2) \ .$$

The corresponding conformal map $f_{\lambda,b}: S^4 \longrightarrow S^4$ is $f_{\lambda,b}(x) = \lambda(b - x)^{-1}$. Put $u(x) = \lambda(\overline{b - x})^{-1}$. Then all instantons in P_1 are given by

$$\mathrm{Im}\left(\frac{\bar{u} \cdot du}{1+|u|^2}\right)$$

where $u(x) = \lambda(\overline{b - x})^{-1}$ and $(\lambda, b) \in R_+^1 \times H$.
<u>Remark:</u> The moduli space $\mathcal{M}^+(P_k)$ $(k > 0)$ has a similar description. In fact, if we replace $\lambda \in R_+^1$ by a vector $\lambda = (\lambda_1, \dots, \lambda_k)$ and $b \in H$ by a $(k \times k)$-matrix, then all SU(2)-instantons in P_k are given by the same formula (see [2] , [5]). However, in this general case we need some new ideas related with the twistor program to prove that one gets <u>all</u> self-dual solutions of the Yang - Mills equations. In the next section we shall explain this twistor space of a 4-dimensional Riemannian manifold.

6. Some remarks and applications of the twistor space.

The main point of the twistor program is that one can translate certain problems of 4-dimensional Riemannian geometry into problems of 3-dimensional complex analysis. To explain this idea let us first of all look at the case of dimension two.

Consider a 2-dimensional oriented Riemannian manifold (M^2,g). Let $J : TM^2 \longrightarrow TM^2$ be the rotation on the angle $\pi/2$ in positive direction. Then J is an integrable complex structure compatible with the metric. Furthermore, two Riemannian metrics g_1 and g_2 on M^2 are conformally equivalent if and only if the corresponding complex structures coincide. Hence in case of dimension two we have a bijective correspondence between conformal structures and complex structures on M^2.

The case of dimension $n = 4$ is more complicate since the set of all complex structures $J : R^4 \longrightarrow R^4$ compartible with the Euclidean structure and the orientation is diffeomorphic to $SO(4)/U(2) = P^1(C) = S^2$. (this set contains more than one point !). If (M^4,g) is an oriented Riemannian manifold, we define

$$P_x = \left\{ J_x : T_xM^4 \longrightarrow T_xM^4 : J_x^2 = -1 \text{ , } J_x \text{ is compatible with the metric and orientation.} \right\}$$

Then $P = \bigcup_{x \in M^4} P_x$ is a $P^1(C)$-fibration over M^4. P is called the twistor space of the Riemannian manifold M^4. Using the Levi-Civita connection on M^4 we split the tangent bundle of P into vertical and horizontal subspaces , $TP = T_vP + T_hP$. We define an almost complex structure $J : TP \longrightarrow TP$ by the conditions:

1.) On vertical vectors J coincides with the canonical complex structure of the fibre $P^1(C)$.
2.) On horizontal vectors at $J_x \in P$ the operator J coincides with J_x.

Theorem (see [1] , [7]): (P,J) is a complex structure if and only if M^4 is a self-dual Riemannian manifold.

Remark: The twistor space of S^4 is $P^3(C)$. Furthermore, if (Q,α) is an instanton on M^4 and $E = Q \times_G V$ is the corresponding vector bundle, then $\pi^*(E) \longrightarrow P$ is a holomorphic vector bundle with some special properties over the twistor space (Ward). It follows that the classification of all SU(2)-instantons on S^4 is equivalent to the classification of rank 2 vector bundles with some addinal structures over $P^3(C)$. This classification problem in algebraic geometry was solved by Drinfeld and Manin [5] and hence one gets the classification of all SU(2)-instantons mentioned in section 5.

Finally we shall explain a geometric application of the twistor space and we sketch the proof of the following

Theorem ([6] , [8]): If M^4 is a closed connected self-dual Einstein manifold with positive scalar curvature, then $M^4 = S^4$, $P^2(C)$.

The proof of this theorem decomposes into three steps:

1. The differential-geometric part:

Consider a four-dimensional oriented Riemannian manifold M^4 and its twistor space P. We introduce a hermitian metric g^λ on P by pulling back the metric of M^4 to the horizontal subspaces and by adding the λ - fold of the metric of the fibres.

Theorem 1: (P , J , g^λ) is a Kähler manifold if and only if M^4 is a self-dual Einstein space of positive scalar curvature $\tau = 48/\lambda > 0$. Furthermore, in this case P becomes an Einstein-Kähler manifold with positive scalar curvature and the base space M^4 is simply connected.

2. The topological part:

Denote by $\gamma \in H^2(P ; Z)$ the first Chern class of the complex vector bundle T_vP. Then $\gamma = c_1(T_vP) = c_1(T_hP)$ and the cohomology ring $H^*(P ; Z) = H^*(M^4 ; Z)[\gamma]$ as well as the characteristic classes of P are given by

$$c_1(P) = 2\gamma \ , \ c_2(P) = 3(\sigma - \chi)\pi^*[M^4] \ , \ c_3(P) = -\gamma\cdot\pi^*[M^4] \ ,$$
$$\gamma^2 = (3\sigma - 2\chi)\pi^*[M^4] \ .$$

In case M^4 is a self-dual Einstein space with positive scalar curvature, using the Bochner theorem ($h^{p,o}(P) = 0$, $p > 0$) and the Hirzebruch-Riemann-Roch formula we obtain

Theorem 2: If M^4 is a self-dual Einstein space of positive scalar curvature, then M^4 is simply connected and the quadratic form $H^2(M^4,Z) \times H^2(M^4,Z) \longrightarrow Z$ is positive definite and $b_2(M^4) = \sigma(M^4) \leq 3$.

In case $b_2 = 0$, M^4 is isometric to S^4.

If $b_2 = 1$, one can prove that M^4 has to be the complex projective plain $P^2(C)$. In the third part we exclude the cases $b_2 = 2$, 3 .

3. The algebraic-geometric part:

Denote $b = b_2(M^4) = \sigma(M^4)$ and consider the exact sequence of sheaves $Z \longrightarrow \mathcal{O} \longrightarrow \mathcal{O}^*$. Then

$$\ldots \longrightarrow H^1(P,\mathcal{O}) \longrightarrow H^1(P,\mathcal{O}^*) \xrightarrow{\partial} H^2(P,Z) \longrightarrow H^2(P,\mathcal{O}) \longrightarrow \ldots$$

is exact and $H^1(P,\mathcal{O}) = 0 = H^2(P,\mathcal{O})$ by Bochner's theorem. This means that $\partial : H^1(P,\mathcal{O}^*) \longrightarrow H^2(P,Z)$ is an isomorphism and hence there exists a unique holomorphic bundle L_γ such that

$\partial L_\gamma = \gamma$. Using the cohomology structure of P described above as well as the Hirzebruch-Riemann-Roch formula we calculate
$\dim H^0(P, L_\gamma) = 9 - 2b$.

<u>Theorem 3:</u> The linear system $/L_\gamma/$ has no base points. Hence there exists a morphism $\phi : P \longrightarrow P^{9-2b}(C)$. If b = 3 , then $\phi(P) = P^3(C)$ and ϕ is a double covering ramified along a K3-surface. If b = 2 , then ϕ is an imbedding and $\phi(P)$ is the intersection of two quadrics. In both cases the Euler characteristic $\chi(P)$ satiesfies $\chi(P) \leq 0$ and hence P cannot be the twistor space of a self-dual Einstein space of positive scalar curvature.
In case b = 1 , P is analytic isomorphic to the flag manifold F(1,2) and the base space has to be $P^2(C)$.

References

[1] M.F.Atiyah,N.J.Hitchin,I.M.Singer,Self-duality in four dimensional Riemannian geometry , Prc.Roy.Soc.Lond. A 362 (1978) , 425 - 461.

[2] M.F.Atiyah, Geometry of Yang-Mills Fields, Pisa 1979.

[3] J.P.Bourguignon,H.B.Lawson,Stability and Isolation phenomena for Yang-Mills fields, to appeat in Comm.Math.Physics.

[4] J.B.Bourguignon,H.B.Lawson,Yang-Mills-Theory - its physical origins and differential geometric aspects,preprint.

[5] V.G.Drinfeld,YuI.Manin,Instantons and sheaves on CP^3, Funkt. Analiz,13 (1979), 59-74.

[6] Th.Friedrich,H.Kurke,Compact four-dimensional self-dual Einstein manifolds with positive scalar curvature,to appear in Math. Nachr.

[7] Self-dual Riemannian Geometry and Instantons, Proceedings of a Summer School on Yang-Mills-Equations held in Kagel 1979, ed. by th. Friedrich, Leipzig 1981.

[8] N.J.Hitchin,Kählerian twistor spaces, to appear in Proc.London Math.Soc.

[9] B.Kostant,Quantization and unitary representations,Lecture Notes in Mathematics 170 (1970), 87-207.

[10] A.S.Schwarz,Instantons and Fermions in the field of Instantons, Comm.Math.Physics 64 (1979), 233-268.

[11] A.Trautman,Solutions of the Maxwell and Yang-Mills Equations associated with Hopf fibrings,International Journ. of Theoretical Physics 16 (1977),561-565.

[12] K.Uhlenbeck,Removable singularities in Yang-Mills fields, Bull.Amer.Math.Soc. 1 (1979),579-581.

SYMMETRY AS A CLUE TO THE PHYSICS OF ELEMENTARY PARTICLES

F.A. Bais

Institute for Theoretical Physics
P.O. Box 80.006, 3508 TA Utrecht
The Netherlands

ABSTRACT

The basic ingredients of the present theories of elementary particles and their interactions are reviewed on an elementary level. The aim is to convey the physical motivations for gauge theories using the notion of symmetry as a guiding principle. Realistic theories like Quantum Electrodynamics, the Glashow-Weinberg-Salam model, and Quantum Chromodynamics are briefly discussed. The idea of Grand Unification is introduced.

TABLE OF CONTENTS

1. ELEMENTARY PARTICLES & FUNDAMENTAL INTERACTIONS

In the history of physics the qualifications "elementary" and "fundamental" have changed substantially over time. What do they mean today?

We distinguish at present four fundamental interactions (forces) between the elementary constituents of matter. Gravity and Electromagnetism (i.e. the unified theory of electricity and magnetism) are known the longest because they are manifest on a macroscopic scale. The other two are the weak interaction(s) responsible for example for the radio-active nuclear decay processes, and the strong interactions causing the (relatively) tight binding of protons and neutrons in the nucleus. For what we know experimentally at present, this list is complete, but it is very well possible that there are others, notably extremely weak interactions which are simply to weak to be observed, alternatively there could be super strong interactions whose effects are limited to an extremely small range.

The elementary constituents on which these forces act are divided to fall into two classes. The first consists of the *leptons* (weakly interacting particles), these are characterized by the fact that they are insensitive to the strong force rather than that they are acted upon by the weak force. The most well known is the electron. Another example is the electron neutrino, the sneaky particle without mass or electric charge subject to the weak interactions and gravity only. The second class is made up of the particles which do feel the strong force: the *quarks*. Ordinary physics, that is the physics of protons, neutrons and electrons which make up all stable nuclei and atoms, involves only two types of quarks (usually denoted as "up" and "down" quark) and the electron and its neutrino. These make up what is called a "family" of elementary particles. Though nobody ordered them,it seems that nature has provided us with at least three of such families. We have no deep understanding of why they are there.

As we will see, all the interactions we mentioned are described by a gauge theory. The notion of a force has a natural interpretation in the context of a gauge theory. It is associated with a special vector field; the *gauge potential* or connection one form. This field is associated with a particle, often denoted as *exchange particle*, carrying the force between the charges it couples to. For example, the familiar Coulomb force between two electric charges can be ascribed to the exchange of a photon between the two charges.

It should be kept in mind that the exchange particles associated with the fundamental forces may be acted upon by these forces as well. For example, the fact that light from distant stars is deflected by the sun, shows that gravity couples to the photon. Yet we like to distinguish the exchange particles from the constituent particles discussed before.

2. PARTICLES AND FIELDS

Particle theory is predominantly expressed in the language of relativistic quantum field theory. This formalism reconciles the two profound and radical ideas that entered physics early on in this century: Special Relativity and Quantum Mechanics. It is beyond the scope of this lecture to even summarize this well established subject (some aspects are discussed in the lecture of P.J.M. Bongaards). To continue our discussion however a few basic facts are indispensable.

In Quantum Field Theory, particles are associated with causal fields. These fields are local operators which act on a multiparticle Hilbert space, thereby creating or annihilating what we mean by a particle. Furthermore these fields form (a basis for) representations of the transformation groups which correspond to the symmetries of nature.

What does it mean to have a theory of elementary particles? In practice it means that somebody has postulated some Lagrange density $\mathcal{L}$ to which the machinery of quantum field theory can be appied. $\mathcal{L}$ is a scalar (density) under the symmetries of nature and expressed (usually as a polynomial) in the fields and their derivatives:

$$\mathcal{L} = \mathcal{L}\left[\phi(x), \partial_\mu \phi(x), \psi_\alpha(x), \partial_\mu \psi_\alpha(x), A_\mu(x), \ldots\ldots\right] . \qquad (2.1)$$

For example a free spin ½ fermion of mass m would be described by

$$\mathcal{L} = \bar{\psi}_\alpha \left\{\gamma^\mu_{\alpha\beta} i \partial_\mu - m\right\} \psi_\beta \qquad (2.2)$$

where ψ_α $(\alpha = 0,\ldots,3)$ is a four component Dirac spinor,

$$\bar{\psi}_\alpha = \left(\psi^\dagger \gamma^0\right)_\alpha \qquad (2.3)$$

and γ^μ are the 4×4 Dirac matrices satisfying

$$\left\{\gamma^\mu,\gamma^\nu\right\} = 2g^{\mu\nu} . \tag{2.4}$$

The action is defined as the spacetime integral of $\mathcal{L}$,

$$I = \int d^4x\, \mathcal{L} . \tag{2.5}$$

The Euler-Lagrange (or field) equations are obtained by imposing stationarity of the action under small variations in the fields. In the example (2.2) one obtains of course the Dirac equation

$$(i\not{\partial}-m)\psi = 0 \qquad \bar{\psi}(i\overleftarrow{\not{\partial}}+m) = 0 \tag{2.6}$$

with $\not{\partial} \equiv \gamma^\mu\partial_\mu$.

The Lagrangian density completely specifies the theory in a very efficient way. It displays the fields which are involved, determines their dynamics, and exhibits most invariances quite explicitly. Not any $\mathcal{L}$ will lead to a consistent theory and there are many restrictions on $\mathcal{L}$ to produce a physically acceptable theory. For example it should generate a definite metric on the space of physical states (this is something like saying that we only allow positive probabilities for physical processes). Free theories like the one defined by (2.2) can be solved exactly. Theories of particles which interact we know in general only how to solve approximately by means of a perturbation expansion in the relevant coupling constants (i.e. interaction parameters). In fact the perturbation series for the realistic models we are about to discuss are at best asymptotic. So if the expansion parameter is small enough this does not pose a problem in practice.

There is another aspect of perturbation theory when applied to fields which we like to mention. It is the problem of renormalization. There are certain physical quantities we know to be finite (like the electron mass or charge for example) which appear to be infinite when calculated naively in perturbation theory. There is a procedure called renormalization by which these apparent infinities are taken care of: they are absorbed in an (infinite) renormalization of the "bare" parameters and fields of the theory (like the mass parameter and the coupling constant). The condition of renormalizability requires that all divergent quantities that may appear to any given order of the perturbation series can be taken care of by a finite number of renormalizations of the parameters and fields in the original

Lagrangian. It imposes severe constraints on the Lagrangian you start of with. For example, in four dimensions it forbids terms in $\mathcal{L}$ in which combinations of fields appear with a combined dimension larger than $(\text{mass})^4$. Though a fundamental justification beyond perturbation theory for the restriction is lacking, it has been of tremendous importance. In the first place, it makes highest order calculations possible, allowing in many cases extremely accurate predictions. Secondly it has been of great use in discovering the theories which are successful in describing the physics at presently attainable energy and distance scales.

3. SYMMETRIES (General)

Symmetry considerations play a crucial role in the design and understanding of the theories of elementary particles. There are however some important distinctions to be kept in mind when talking about symmetry.

In the first place there is a deep physical distinction between *global* and *local* symmetries. A global symmetry is one where the *same* transformation should be applied at every point in space time. In the case of a local symmetry the transformation may differ from point to point in a continuous way.

Secondly, there is a distinction between the ways in which a symmetry can be realized in the physical states. It may be realized *manifestly* or *hidden*. In the latter case we speak of a *broken* symmetry.

Finally there is the important physical fact that many symmetries in nature are only approximate. One may imagine a small term in the Lagrangian which is not invariant whereas the others are. In such a case still a lot can be learned from symmetry considerations. An intriguing fact is that all exact symmetries we know of in nature are local symmetries. It is not known why this is so.

4. GLOBAL SYMMETRIES

The fields appearing in the Lagrangian form a basis for representations of the various symmetry groups which that Lagrangian may have. We should thus think of the fields corresponding to elementary particles as carrying many types of indices indicating how they transform under which group.

Applying a transformation to the fields (the index i stands for all indices which are *not* affected)

$$\psi_i \rightarrow \psi_i' \tag{4.1}$$

leaves the Lagrangian invariant, i.e. infinitesimally one has

$$\mathcal{L} \rightarrow \mathcal{L}' = \mathcal{L} + \delta\mathcal{L} \ , \ \delta\mathcal{L} = 0 \ . \tag{4.2}$$

In quantum language we would say that since the symmetry operators commute with the Hamiltonian which governs the time evolution of the system, we may label the physical states by the eigenvalues of some maximal number of mutually commuting symmetry operators. Indeed it would be hard to characterize a physical system if there were no symmetries.

A most important consequence of a symmetry of the Lagrangian is the existence of a conserved charge (Noether's theorem). The change in the Lagrangian under some infinitesimal transformation can be written as a total divergence

$$\partial\mathcal{L} = \partial_\mu j^\mu \tag{4.3}$$

where the current j_μ is of the form

$$j^\mu = iq \sum_i \frac{\delta\mathcal{L}}{\delta(\partial_\mu\psi_i)} \psi_i \ . \tag{4.4}$$

The vanishing of $\delta\mathcal{L}$ implies that the charge

$$Q = \int d^3x \, j^o \tag{4.5}$$

is conserved (assuming vanishing current through the boundary):

$$\partial_t Q = 0 \ . \tag{4.6}$$

For example the Lagrangian (2.2) has a U_1 phase invariance

$$\psi \to \psi' = e^{ie\alpha}\psi \tag{4.7}$$

so that the current would just be

$$j^\mu = e\bar{\psi}\gamma^\mu\psi . \tag{4.8}$$

Similarly one may derive the conservation of energy momentum and angular momentum as a consequence of the invariance of the Lagrangian under the inhomogeneous Lorentz group.

5. LOCAL SYMMETRIES (Abelian)

Let us now consider local symmetries, where the transformation parameter is allowed to depend on the space time coordinates. We may ask what we have to change in the Lagrangian (2.2) in order to make it invariant under local transformations

$$\psi \to \psi' = e^{i\alpha(x)}\psi . \tag{5.1}$$

$\mathcal{L}$ is not invariant because

$$\partial_\mu\psi \to \partial_\mu\psi' = e^{i\alpha(x)}(\partial_\mu + i\partial_\mu\alpha)\psi . \tag{5.2}$$

To remedy the situation we have to introduce a *gauge potential* (or *connection one-form*) A_μ and a *covariant derivative*

$$D_\mu\psi \equiv (\partial_\mu - ieA_\mu)\psi . \tag{5.3}$$

For the Lagrangian with ordinary derivatives replaced by covariant ones to be invariant, the following transformation properties suggest themselves

$$D_\mu\psi \to D_\mu\psi' \equiv e^{i\alpha} D_\mu\psi \tag{5.4}$$

and

$$A_\mu \to A'_\mu = A_\mu + \frac{1}{e} \partial_\mu \alpha \ . \tag{5.5}$$

In order to have an internal symmetry of the kind described it is apparently necessary to introduce an additional local vector field.

At this point however the field A_μ plays a rather uninteresting role. Its field equation is trivial and could be used to eliminate A_μ from the Lagrangian altogether, in favor of a complicated interaction between the other fields. To endow the field A_μ with physical degrees of freedom, i.e. associate with it a particle that propagates in space time, one has to add an invariant term involving the derivatives of A_μ to $\mathcal{L}$. At this point the *Field strength* (or *curvature*)

$$F_{\mu\nu} = \partial_\mu A_\nu - \partial_\nu A_\mu \tag{5.6}$$

enters the physical picture. It is just the electromagnetic field (or Maxwell) tensor with

$$\begin{aligned} F_{ij} &= \varepsilon_{ijk} B_k \\ F_{oi} &= E_i \end{aligned} \tag{5.7}$$

where $\vec{B}$ and $\vec{E}$ are the magnetic and electric field respectively.

It is gratifying to see that even in this unrefined approach the fundamental concepts of differential geometry such as connection, covariant derivative and curvature, are forced upon us as the basic ingredients of the physical theory. The reader may have realized that in the mean time we have gathered all items needed to construct the Lagrangian of Quantum Electrodynamics the prototype of realistic gauge theories.

6. QUANTUM ELECTRODYNAMICS (QED)

The simplest Lagrangian with a local U_1 invariance is of the form

$$\mathcal{L} = -\tfrac{1}{4} F_{\mu\nu}^2 + \bar{\psi}(i\not{D}-m)\psi . \tag{6.1}$$

No other (higher order) invariants can be added if one is to require the theory to be renormalizable. The theory describes the abelian gauge field A_μ [the photon] interacting with a massive charged fermion with spin ½ [for instance the electron and its antiparticle]. The interaction term [hidden in the covariant derivative] takes the form

$$A_\mu j_\mu = e\bar{\psi}\ \gamma_\mu \psi\ A_\mu \tag{6.2}$$

where j_μ is the conserved electromagnetic current (4.8). The gauge particle evidently couples to a conserved current in a way which is completely determined by the local gauge symmetry. Note that a mass term for A_μ which would be of the form

$$\Delta\mathcal{L} = -\mu^2\ A_\mu A^\mu \tag{6.3}$$

is not invariant under the transformation (5.5) and hence excluded. The fact that A_μ is necessarily massless implies that the force it mediates has an infinite range.

The field equations are

$$\begin{aligned} \partial_\mu F^{\mu\nu} &= e\bar{\psi}\ \gamma^\mu\ \psi \\ (i\not{\partial}-m)\psi &= e\not{A}\psi \end{aligned} \tag{6.4}$$

a complicated nonlinear system for operator valued quantum fields. The only way we know how to solve them, is approximately by means of perturbation theory. We start with the solutions for the free fields, i.e. we set the right hand side of (4.4) equal to zero, and iterate the solution. The expansion parameter is the "fine structure constant" $\alpha = e^2/4\pi \simeq 1/137$. Its smallness guarantees a rapid convergence. QED is one of the best confirmed theories of physics. It has been tested over a wide range of distances varying from inside the proton to galactic separations. The theoretical prediction of the anomalous magnetic moment of the electron coincides with the experimental value up to eleven significant decimal places.

7. LOCAL SYMMETRY (Non abelian)

What happens if the local symmetry group G is non abelian? This possibility was first considered in the context of particle physics by Yang and Mills, and Shaw (1954). In the present context we are only interested in cases where G is compact (semi simple), because only then is it possible to construct a positive definite energy density from the Lagrangian. This is a consequence of the fact that the Killing form of the (real valued) Lie algebra has a definite sign. The fields transform under finite dimensional unitary transformations $\Omega \in G$ as

$$\psi' = \Omega\psi \tag{7.1}$$

where

$$\Omega = \exp(i\alpha^a(x)T_a) \ . \tag{7.2}$$

The parameters $\alpha^a(x)$ $(a = 1,2,\ldots,\dim G)$ are real. The T_a form a representation of the (hermitean) generators of the group G, satisfying the standard commutation relations

$$\left[T_a, T_b\right] = if_{abc}\, T_c \ . \tag{7.3}$$

The structure constants f_{abc} are totally antisymmetric.

The Lagrangian is constructed in terms of covariant quantities. The covariant derivative takes the form (5.3). The gauge potential A_μ is the (hermitean) Lie algebra valued one form

$$A_\mu \equiv A_\mu^a\, T_a \tag{7.4}$$

which transforms as

$$A_\mu \to A'_\mu = \Omega A_\mu \Omega^{-1} - \frac{i}{e}(\partial_\mu\Omega)\Omega^{-1} \ . \tag{7.5}$$

The fieldstrength (curvature) takes the form

$$F_{\mu\nu} = \partial_\mu A_\nu - \partial_\nu A_\mu + i[A_\mu, A_\nu] \tag{7.6}$$

and has acquired an extra term nonlinear in A_μ. At this point we are in a position to present the Lagrangian of Quantum Chromodynamics, a theory which is widely believed to correctly describe the strong interactions between the quarks. Its Lagrangian is in fact as simple as a non abelian gauge (Yang Mills) theory can be.

8. QUANTUM CHROMODYNAMICS (QCD)

The color gauge group is $G = SU_3^c$. Quarks form a triplet, i.e. a basis for the defining representation of SU_3^c. The Lagrangian density is

$$\mathcal{L} = -\tfrac{1}{4} F^a_{\mu\nu} F^a_{\mu\nu} + \sum_i \bar{\psi}_i (i\not{D} - m_i)\psi_i \qquad (8.1)$$

There are 8 [= dim SU_3] gauge particles A^a_μ called "gluons". The summation in the second term is over the various species of quarks which are necessary to build all forms of strongly interacting matter we know of. At present there are five species (up, down, strange, charm and bottom) and the discovery of a sixth kind (top) is expected. Each of these form a color triplet.

Despite the similar appearance of (8.1) and the Lagrangian (6.1) of the abelian theory, there are important differences which make the non abelian theory much more intricate. The situation with QCD at present resembles that, once described by Galileo, of a block of marble which contains a beautiful statue or perhaps many. The problem is to reveal them.

The essential complication of the theory resides in the fact that it is intrinsically nonlinear already at the level where one ignores the quarks. This is manifest in the expression (7.6) for the curvature. It expresses the physical fact that the gluons themselves carry color charge (also clear from the fact that they transform homogeneously as well in (7.5)), this in contrast with the photon which is electrically neutral.

Even after Gell-Mann and Zweig proposed the quark idea (1963) it took about a decade before the applicability of the non abelian gauge theory to the problem of the strong interactions was realized. (It took a lot of effort to overcome our abelian intuition and prejudices.)

The main obstacle to a direct empirical verification of the theory is the phenomenon of confinement. Quarks cannot be observed as isolated free particles. They only appear in combinations which are color singlets, like *mesons* (such as the pion) consisting of a quark and an antiquark and *baryons* (such as the proton and neutron) consisting of three quarks.

What does perturbation theory teach us? The naive first order approximation suggests massless gluons mediating some Coulomb type interaction between isolated quarks. Rather misleading indeed. In the next order one discovers something extremely interesting. The (renormalized) effective coupling constant has a dependence on momenta q^2 (or inverse distance $1/r^2$) which takes the form

$$e^2(q^2) = \frac{e_o^2}{1-b\ \ell n(q^2/\Lambda^2)} \tag{8.2}$$

where

$$b = 11 - \frac{2}{3} N_F \tag{8.3}$$

with N_F = number of species $\cong 5$.
Apparently if we probe a strongly interacting system with very high energies (measured in terms of some parameter Λ) i.e. $q^2 \to \infty$, or equivalently at very small distances, then $e^2(q^2) \to o$ and the theory behaves like a free theory. This property is called asymptotic freedom. It means that short distance phenomena can be treated successfully with perturbation theory. At the other hand for decreasing q^2 the effective coupling constant grows (infrared slavery) and becomes so large as to invalidate the perturbative argument on which formula (8.2) was based. Perturbation theory necessarily breaks down beyond a certain scale. Consequently, the asymptotic states of the theory may be very different from what perturbation theory suggests. For example quarks may be confined.

It is at this point that semiclassical considerations entered the discussion. It motivated the interest in the topological nontrivial structure of gauge theories, also on the classical level, initiated by 't Hooft and Polyakov. A complete understanding of the classical solutions of the pure Yang-Mills equations

$$D_\mu F_{\mu\nu} = o$$

is still lacking. Much progress has been made (as will be the subject of most of the lectures at this meeting) on the problem of the (anti) selfdual gauge fields. These satisfy

$$F_{\mu\nu} = \pm \tilde{F}_{\mu\nu}$$

where the dual of the fieldstrength is defined by

$$\tilde{F}_{\mu\nu} = \tfrac{1}{2}\, \varepsilon_{\mu\nu\alpha\beta}\, F_{\alpha\beta} \ .$$

These are automatically solutions of the field equations due to the Bianchi identity $D_\mu \tilde{F}_{\mu\nu} \equiv 0$. The role these selfdual solutions play in the quantum theory is still poorly understood, though in the semi-classical approximation some interesting properties of the theory have been demonstrated.

9. HIDDEN GLOBAL SYMMETRIES

The fact that the dynamics of a physical system i.e. that the Lagrangian or field equation has a certain symmetry, does not imply that this symmetry is manifest in the groundstate of the system. Looking at the inside of a magnet it is far from obvious that the equations governing the interactions of the spins which make up the overall magnetization are rotationally symmetric. We speak of a hidden or alternatively of a (spontaneously) broken symmetry. It is a statement about the groundstate rather than about the dynamics, so a "broken" symmetry can be "exact". We prefer the term hidden symmetry because it implies that it is rather a different realization of symmetry than the absence of symmetry. This idea has many important applications in (particle) physics.

Consider a Lagrangian of a complex scalar field with a global U_1 invariance

$$\mathcal{L} = -\ (\partial_\mu\phi)^\dagger\ (\partial_\mu\phi) - V(\phi^\dagger\phi) \tag{9.1}$$

We choose the potential V, which describes the selfinteractions of the scalar field of the form

$$V(\phi^\dagger\phi) = \frac{\lambda}{4}\ (\phi^\dagger\phi - f^2) \tag{9.2}$$

with f and λ real constants and furthermore $\lambda \ll 1$.
The groundstate (to lowest order in λ) is one where ϕ acquires a non-vanishing expectation value which equals a value which minimizes the classical potential

$$|\langle\phi\rangle| = |f| \; . \tag{9.3}$$

It is in this sense that the groundstate of the theory "breaks" the U_1 invariance of $\mathcal{L}$. To exhibit the physical implications we expand the field ϕ around a value in the degenerate groundstate

$$\phi = f + \frac{1}{\sqrt{2}} \left\{ \chi(x) + i\sigma(x) \right\} \tag{9.4}$$

where χ and σ are both real fields. Substitution of (9.4) in $\mathcal{L}$ and keeping only the lowest order term in λ yields

$$\mathcal{L} \simeq - \tfrac{1}{2}(\partial_\mu \chi)^2 - \tfrac{1}{2}(\partial_\mu \sigma)^2 - \tfrac{1}{2}\lambda f^2 \chi^2 + \ldots \tag{9.5}$$

Apparently this theory describes a field χ with mass $m_\chi = \sqrt{\lambda}\, f$ and a field with vanishing mass $m_\sigma = o$. The χ field is associated with the radial oscillations whereas the σ-field is related with the angular oscillations where the potential is flat. The massless particle is usually called the (Nambu) Goldstone particle. It is a consequence of the vacuum degeneracy which arises if one breaks a continuous global symmetry by some expectation value of the fields. The number of massless excitations in such a groundstate equals the dimension of the vacuum manifold.

The classical example is the ferromagnet which we mentioned before. In particle physics the pion particles are interpreted as the Goldstone particles of a spontaneously broken non abelian chiral symmetry in the Lagrangian of QCD. The pions acquire a small mass because the chiral symmetry is only approximate in the first place (it is violated by the explicit mass terms of the quarks).

10. HIDDEN LOCAL SYMMETRIES

Using the recipe given in Sect. 5 we turn the global U_1 symmetry of (9.1) into a local one by replacing the ordinary derivatives by covariant derivatives. What is the effect of the vacuum expectation value ϕ? Observe that if

$$\phi = \left[f+\chi(x)\right]e^{-i\sigma(x)} \tag{10.1}$$

the covariant derivative may be expanded as

$$D_\mu\phi = D_\mu\chi - ief\left(A_\mu + \frac{1}{e}\,\partial_\mu\sigma\right) + \ldots \tag{10.2}$$

This invites us to consider the field

$$A'_\mu = \left(A_\mu + \frac{1}{e}\,\partial_\mu\sigma\right) \tag{10.3}$$

which is just a gauge transformation of A_μ. $F_{\mu\nu}$ may as well be thought of as the curl of A'_μ. To lowest order the relevant term in the Lagrangian takes the form

$$|D_\mu\phi|^2 = (D_\mu\chi)^2 + e^2f^2A'^2_\mu\ . \tag{10.4}$$

Surprise! The massless σ-particle has disappeared altogether and instead we have generated a mass for the gauge particle $m_A = ef$. The phrase which goes with this so called Higgs effect is that "the gauge particle has acquired a mass by eating the massless Goldstone particle". This prose is not as inaccurate as one may think. In the sense of counting the degree's of freedom this is basically what happens. A massless vector particle has only two (transverse) polarization states, whereas a massive vector particle has three. The third (longitudinal) polarization state is just the degree of freedom associated with the σ-field.

Whereas the mathematics of the Higgs effect is utterly trivial the physics is not. We have shown the way a mass for the gauge particles may be generated without destroying the local symmetry. It is only after the discovery of this possibility that the application of the non abelian gauge theory to the problem of the weak interactions became feasible. As was proved by 't Hooft in 1971 it is only by introducing a mass this way that a non abelian theory remains renormalizable. We note that the orientation of the vacuum expectation value is not gauge invariant and hence is not observable. The symmetry is realized in a way characterized by the massive gauge particle (i.e. short range interactions) and the appearance of topological excitations as will be the subject of the next lecture.

The canonical example in physics is the superconductor. The Lagrangian we described is just the relativistic generalization of the Landau-Ginzburg theory of superconductivity (also named abelian Higgs model). The vacuum expectation value of the scalar field describes the bose condensation of Cooper pairs in the superconducting ground state. The mass of the gauge field manifests itself through the Meissner effect: the fact that a magnetic field is expelled from a superconducting region.

How does the Higgs mechanism work in non abelian theories? This we will illustrate by presenting an abridged version of the Glashow-Weinberg-Salam theory of the weak and electromagnetic interactions (1967) which by now has successfully survived an extensive confrontation with experimental data.

11. THE GLASHOW-WEINBERG-SALAM THEORY

The weak force is not just characterized by its weakness but also by the fact that it acts only over very small distances ($\leqslant 10^{-15}$ cm). It suggests that the weak exchange particle(s) have a mass μ because that gives rise to a Ukawa potential

$$V(r) = G \frac{e^{-\mu r}}{r} \tag{11.1}$$

between two weakly charged particles with a separation r. The range of the force is then of order μ^{-1}. The present unified theory of the weak and electromagnetic interactions is a gauge theory based on the group $G = SU_2^L \times U_1^Y$ where the weak gauge particles are given a mass through the Higgs mechanism. The theory has three gauge fields W_μ^a $(a = 1,2,3)$ associated with the generators T_a of SU_2^L and one gauge field B_μ corresponding to the U_1 generator Y of "weak hypercharge". We restrict ourselves to introducing the electron and its neutrino as the elementary fermions. The complete theory also features the other leptons with their neutrinos and the quarks. Their weak interaction properties are actually very similar.

A fact that shocked many at the time of its discovery is that the weak interactions violate parity symmetry. This lack of mirror symmetry in nature is reflected in the fact that the weak gauge particles couple only to the left handed components of the fundamental fermions. We split the fermion field in their left and right handed components

$$\psi = \psi_L + \psi_R \tag{11.2}$$

with

$$\begin{aligned} \psi_L &= \tfrac{1}{2}(1-\gamma_5)\psi \\ \psi_R &= \tfrac{1}{2}(1+\gamma_5)\psi \end{aligned} \tag{11.3}$$

(γ_5 is the special Dirac matrix $\gamma_5 = i\gamma^0\gamma^1\gamma^2\gamma^3$).
The left handed components form a doublet L under SU_2^L

$$L = \begin{pmatrix} \psi_\nu \\ \psi_e \end{pmatrix}_L , \tag{11.4}$$

whereas the right-handed component of the electron,

$$R = (\psi_e)_R ,$$

is a singlet. The neutrino only has a left-handed component (this is possible because it is assumed to be massless).
The electric charge operator

$$Q = T_3 + \tfrac{1}{2}Y \tag{11.5}$$

yields the correct charges if $Y_L = -1$ and $Y_R = -2$.

The part of the Lagrangian describing the gauge fields is

$$\mathcal{L}_{gauge} = -\tfrac{1}{4} G^a_{\mu\nu}G^a_{\mu\nu} - \tfrac{1}{4} f_{\mu\nu}f_{\mu\nu} \tag{11.6}$$

where $G^a_{\mu\nu}$ and $f^a_{\mu\nu}$ are the SU_2^L and U_1^Y fieldstrengths. The part describing the fermions and their gauge interactions takes the form

$$\begin{aligned} \mathcal{L}_{fermions} &= \bar{R}\, i\gamma^\mu\left(\partial_\mu - \frac{i}{2}\, g' B_\mu Y\right)R \\ &+ \bar{L}\, i\gamma^\mu\left(\partial_\mu - \frac{i}{2}\, g' B_\mu Y - \frac{i}{2}\, g W^a_\mu \tau^a\right)L . \end{aligned} \tag{11.7}$$

We have the freedom to introduce two independent coupling constants g and g' because of the direct product structure of G. The generators T^a in the defining representation are $\frac{\tau_a}{2}$ with τ a traceless hermitean Pauli matrices.

Next we give mass to three of the four gauge bosons so that only the photon survives as a massless particle. To that end we introduce a potential for a complex doublet ϕ (with $Y_\phi = 1$)

$$\phi = \begin{pmatrix} \phi^+ \\ \phi^o \end{pmatrix} . \tag{11.8}$$

We take

$$\mathcal{L}_{Higss} = -(D_\mu\phi)^\dagger (D_\mu\phi) - V(\phi^\dagger\phi) , \tag{11.9}$$

with

$$V(\phi^\dagger\phi) = -\frac{\mu^2}{2}\phi^\dagger\phi + \frac{\lambda}{4}(\phi^\dagger\phi)^2 . \tag{11.10}$$

Minimizing this potential yields

$$|<\phi>| = \frac{v}{\sqrt{2}} = \mu^2/\sqrt{\lambda} . \tag{11.11}$$

And by choosing for example

$$<\phi> = \frac{1}{\sqrt{2}}\begin{pmatrix} o \\ v \end{pmatrix} \tag{11.12}$$

one immediately sees that both SU_2^L and U_1^Y are broken but that the combined generator Q of (11.5) is not (i.e. it leaves (11.12) invariant). This is the electric charge generator and apparently we have broken the $G = SU_2^L \times U_1^Y$ symmetry down to the U_1^γ of electromagnetism. Let us go one step further and consider the resulting mass spectrum. Expanding the ϕ field as

$$\phi = \frac{1}{\sqrt{2}}\begin{pmatrix} o \\ v+\eta(x) \end{pmatrix} \tag{11.13}$$

one obtains,

$$\mathcal{L}_{Higgs} = -\tfrac{1}{2}(\partial_\mu\eta)^2 - \mu^2\eta^2 - $$
$$- \frac{v^2}{4}\left\{\left(g'B_\mu - gW_\mu^3\right)^2 + g^2\left[\left(W_\mu^1\right)^2 + \left(W_\mu^2\right)^2\right]\right\} + \dots \tag{11.14}$$

The following picture emerges. There are two electrically charged massive gauge particles

$$W_\mu^\pm \equiv \frac{1}{\sqrt{2}}\left(W_\mu^1 \mp W_\mu^2\right) \quad , \; m_W = \frac{gv}{2} \; , \tag{11.15}$$

an electrically neutral massive gauge particle

$$Z_\mu^o = \frac{g'B_\mu - gW_\mu^3}{(g^2+g'^2)^{\frac{1}{2}}} \; , \; m_{Z^o} = \sqrt{g^2+g'^2}\,\frac{v}{2} \; , \tag{11.16}$$

and the neutral photon which remains massless

$$A_\mu = \frac{g'B_\mu + g'W_\mu^3}{(g^2+g'^2)^{\frac{1}{2}}} \; , \; m_A = o \; . \tag{11.17}$$

In addition there is a neutral massive scalar particle (η) which survives as a physical particle. The other three degrees of freedom of the original complex scalar doublet ϕ have been transferred to the $W^\pm$ and Z^o particles. Though the $W^\pm$ and Z^o have never been observed directly (they are too massive to be produced in the laboratory as free particles), there is a rather overwhelming indirect evidence for their existence.

We mention that a mass for the fermions can be generated by the Higgs mechanism as well if one adds an interaction between the fermions ψ and the scalar field ϕ.

$$\mathcal{L}_{\phi\psi} = - G_e(\bar{R}\phi^+ L + \bar{L}\phi R) \; . \tag{11.18}$$

This interaction can only generate a mass for the electron but its magnitude is proportional to the merely introduced free parameter G_e. The theory does not predict a relation between the fermion and gauge particle masses.

12. GRAND UNIFICATION

As local gauge symmetry appears to be the underlying principle of all fundamental interactions it is not surprising that attempts have been undertaken to unify QCD and the Glashow-Weinberg-Salam theory (together referred to as the "Standard Model") in one Grand Unified Theory (GUT) with a single (simple) gauge group G. Since G contains the other groups as subgroups

$$G \supset SU_3^c \times SU_2^L \times U_1^Y \tag{12.1}$$

it should at least have rank 4. Suitable candidates are SU_5, SU_{10}, E_6 etc.

We illustrate the main idea in the simplest SU_5 model proposed by Georgi and Glashow (1973). Two stages of symmetry breaking are needed:

$$\underset{10^{14}\mathrm{GeV}}{SU_5 \longrightarrow} SU_3^c \times SU_2^L \times \underset{10^{2}\mathrm{GeV}}{U_1^Y \longrightarrow} SU_3^c \times U_1^Y \; . \tag{12.2}$$

SU_5 has 24 generators which decompose under the $SU_3^c \times SU_2^L$ subgroup as

$$\begin{array}{ccccccccccc} \underline{24} \longrightarrow & (\underline{8},\underline{1}) & + & (\underline{1},\underline{3}) & + & (\underline{1},\underline{1}) & + & (\underline{3},\underline{2}) & + & (\underline{3}^*,\underline{2}) & \\ & \updownarrow & & \updownarrow & & \updownarrow & & \updownarrow & & \updownarrow & . \\ & \text{Gluons} & & W_\mu^a & & B_\mu & & \multicolumn{3}{c}{X,Y \text{ bosons}} & \end{array} \tag{12.3}$$

Twelve of them are readily identified as the gauge particles of the standard model. The others, the so called X and Y bosons transform non trivially under both SU_3^c and SU_2^L. This has important consequences as we will see shortly. Each family of constituent particles (we recall that all "ordinary" physics only involves one family consisting of the up and down quark and the electron with its neutrino) transforms under a reducible $5^* + 10$ representation:

$$\begin{aligned} &\psi^\alpha : \underline{5} \longrightarrow \underset{q_R}{(\underline{3}^*,\underline{1})} + \underset{L}{(\underline{1},\underline{2})} \\ &\psi\{\alpha,\beta\} : \underline{10} \longrightarrow \underset{q_L}{(\underline{3},\underline{2})} + \underset{q_R}{(\underline{3}^*,\underline{1})} + \underset{R}{(\underline{1},\underline{1})} \; . \end{aligned} \tag{12.4}$$

The decomposition allows indeed for the correct quantum numbers for the leptons and quarks.

What is gained by the grand unification of the standard model? We list a few salient features:

1) The U_1^Y group being a subgroup of SU(5) is now compact. Consequently electric charge is quantized. In particular the charge of the electron is exactly equal and opposite to the charge of the proton (i.e. three quarks).
2) The ratio g'/g arbitrary in the GWS theory, is calculable, and agrees with experiment.
3) A rather startling prediction generic for GUTs is caused by the existence of the X,Y bosons. Their quantum numbers imply that they mediate transitions from quarks to leptons. As a consequence, the proton, the basic building block of all matter, is unstable! It can decay for example into a pion and anti-electron. The decay time is directly related with the mass of the X,Y bosons. Since the proton is known to have a lifetime $\tau > 10^{30}$ years, one expects the $m_{X,Y} \gtrsim 10^{14}$ GeV. Obviously the universe has still some time to go. An interesting fact is that if the proton does indeed decay in $\tau < 10^{33}$ years, it will be possible to detect it in especially designed experiments.
4) If GUTs turn out to be correct then all physics up to extremely high energies (10^{15} GeV) would be determined. This is an extrapolation of some thirteen orders of magnitude above the presently attainable energies. The only place and time where such energies may have occurred are in the very early universe ($t \lesssim 10^{-35}$ seconds). This suggests that the properties of GUTs may have left traces in the universe. An exciting instance is in fact the otherwise mysterious abundance of matter over anti-matter in the present universe, which has an explanation in the combined framework of Grand Unification and the standard Big Bang scenario of the universe.
5) We mention that in many GUTs there is no special reason to have exactly massless neutrino's, thereby challenging the traditional view.

13. EPILOGUE

In the previous pages we have roughly sketched the picture that physicists have today of the elementary constituents of matter and their interactions. Models which we believe to be realistic at present were introduced with a heavy emphasis on the role of symmetry arguments. If the reader is left with the impression that physics is close to its completion then that is merely a reflection of the fact that we have refrained from spelling out the shortcomings and problems even of the standard model. These seem sufficient to still feed some generations to come.

The reader will have noticed that the fourth (or first) fundamental interaction, Gravity, has been consistently left out of the discussion. The problem is that though gravity has some properties in common with the other gauge theories, its quantization is a deep and unsolved problem. A conceivable answer may be found along the line of supergravity. N = 8 supergravity is conceivably the theory in which all four fundamental are truly unified. At present this ultimate achievement is far from evident however.

REFERENCES

Instead of the original papers we list some of the many available reviews on the subject.

An introduction to the physical meaning of gauge invariance:

G. 't Hooft, Scientific American.

A detailed discussion of the quantization of gauge fields as well as the GWS theory:

E.S. Abers and B.W. Lee, Phys. Rep. 9 (1973) 1.

J.C. Taylor, Gauge Theories of Weak Interactions (Cambridge Univ. Press, Cambridge, 1976).

QCD:

W. Marciano and H. Pagels, Phys. Rep. 36 (1978) 137.

Grand Unification:

P. Langacker, Phys. Rep. 72 (1981) 185.

Super Gravity:

P. van Nieuwenhuizen, Phys. Rep. 68 (1981) 189.

TOPOLOGICAL EXCITATIONS IN GAUGE THEORIES; AN INTRODUCTION FROM THE PHYSICAL POINT OF VIEW

F.A. Bais

Institute for Theoretical Physics
University of Utrecht
P.O. Box 80.006, 3508 TA Utrecht
The Netherlands

ABSTRACT

Some topological aspects of gauge theories are reviewed with an emphasis on physical intuition rather than mathematical rigor. Magnetic fluxtubes of various sorts (including those with nonabelian flux quantum numbers) are discussed. The basic features of magnetic monopoles in gauge theories are summarized. Besides the standard 't Hooft-Polyakov-monopole we discuss their classification for the general group G, and their fate after crossing a phase boundary. We conclude with a brief look at instantons.

TABLE OF CONTENTS

1. SOLITONS AND INSTANTONS

During the past decennium considerable progress towards unveiling the topological structure of gauge theories has been made[1). At the one hand the question of constructing classical solutions to the non-linear field equations carrying a topological charge has been raised. This problem has attracted the interest of physicists and mathematicians alike and hopefully this lecture will contribute to the ongoing exchange of ideas and techniques in this area. At the other hand the physical implications of the presence of topological charges has been the subject of intensive research.
Important properties of gauge theories (hard, or even impossible to obtain in the conventional perturbative approach) have been revealed. We informally discuss some of the highlights, where the selection of topics reflects to a certain extent the author's preference and limitations.

The importance of classical solutions for the quantum theory is mainly restricted to those cases where the values of the parameters in the theory allow a well defined semiclassical approximation (usually the region of small coupling constants). In certain cases however more general qualitative conclusions about the quantum theory can be drawn from the mere existence of topological charges. The use of a classical solution can be twofold. The distinction between a "soliton" and an "instanton" refers rather to this use than to the type of solution. A soliton (or lump) is a stable solution of finite spacial extend. It is a configuration with finite energy. An instanton (or pseudo particle) is localized both in space and in (Euclidean) time. It is a configuration with finite (Euclidean) action.

Solitons have a clear interpretation as a particle-like object corresponding to some stable bound state of the elementary fields. The classical values of the fields are the expectation values of the corresponding operators in the (lowest approximation of the) quantum state labeled by the appropriate quantum numbers.

The same solution may be interpreted as an instanton in an Euclidean space time of one dimension less. Their role in the quantum theory is more subtle. Instantons give the leading semiclassical contribution to certain vacuum to vacuum amplitudes which would vanish in ordinary perturbation theory. They describe tunneling phenomena which are strictly forbidden classically but can have exponentially small probability amplitudes in the quantum theory. This is most conveniently phrased in the framework of functional integrals. Greens functions are calculated by integrating over classical configurations in Euclidean space time with a measure

$$e^{-S(A)}DA$$

where A denoted the collection of fields and S(A) is the (positive) Euclidean action. Green's functions in Minkowski space with the correct causality properties are then obtained by analytic continuation of the result. Imagine the vacuum (groundstate) to have a discrete degeneracy. A vacuum to vacuum amplitude is computed by performing the integral over all classical configurations which interpolate between particular vacua at $x_4 = -\infty$ and $x_4 = +\infty$. Instantons correspond to paths of finite action in field configuration space connecting the various components of the vacuum. Such tunneling processes (although exponentially small) are important because they may lift the degeneracy of the ground state. A new ground state which is stable under tunneling has to be constructed and may have very different properties. Notably symmetries which were thought to be present may be broken and vice versa.

2. THE ABELIAN HIGGS MODEL

To set the stage for the discussion of topological excitations in gauge theories we turn to a simple example: The Abelian Higgs Model (i.e. the relativistic generalisation of the phenomenological model of Landau and Ginzburg for the superconductor[2]). It features many of the essential aspects of a theory with nontrivial topology. This model, where the local U_1 symmetry is spontaneously broken by a vacuum expectation value of a complex scalar field was discussed to some extend in the previous lecture (Hidden Local Symmetries). The Lagrangian is

$$\mathcal{L} = -\tfrac{1}{4} F_{\mu\nu}^2 + (D_\mu\Phi)^\dagger (D_\mu\Phi) - V(\Phi^\dagger\Phi) , \tag{2.1}$$

with the field strength

$$F_{\mu\nu} = \partial_\mu A_\nu - \partial_\nu A_\mu , \tag{2.2}$$

covariant derivative

$$D_\mu\Phi = (\partial_\mu + ieA_\mu)\Phi \tag{2.3}$$

and a potential of the form

$$V(\Phi^{+}\Phi) = \frac{\lambda}{4} (\Phi^{+}\Phi - f^{2}) \ . \tag{2.4}$$

The U_1 group acts on the fields as follows,

$$\begin{aligned} \Phi &\to \Phi' = e^{i\alpha(x)}\Phi \ , \\ A_{\mu} &\to A'_{\mu} = A_{\mu} - \frac{1}{e}\partial_{\mu}\alpha \ . \end{aligned} \tag{2.5}$$

The potential is minimized for any Φ of the form

$$\Phi = f\, e^{i\alpha} \ . \tag{2.6}$$

In other words the vacuum manifold has the topology of a circle which we denote by S^1_{Φ}. It is obtained as the orbit of some $|\Phi| = f$ under the action of the group (we will ignore the possibility of additional disconnected "accidental" degeneracies in the potential). The theory describes a gauge field with mass $m_A = ef$, and a real scalar field χ with mass $m_{\chi} = \sqrt{\lambda}\, f$ (at least if the space time dimension > 2).

We want to investigate classical solutions which only depend on the (x_1, x_2) or (ρ, ϕ) coordinates, i.e. we consider the model in R^2. It is interesting to study the boundary condition that follows from requiring the energy density $\mathcal{H}(= -\mathcal{L})$ of the two dimensional configuration to be integrable at $\rho \to \infty$. $\mathcal{H}$ is the sum of three positive terms. Integrability implies that each of the terms should go like $\rho^{-3-\varepsilon}$ $(\varepsilon > 0)$. So,

$$\lim_{\rho\to\infty} V(\Phi^{+}\Phi) \simeq 0 \quad \Longrightarrow \quad \Phi(\rho\to\infty) = f\, e^{i\alpha(\phi)} \ , \tag{2.7}$$

where in order for Φ to be single valued we have the additional condition on $\alpha(\phi)$ that

$$\alpha(2\pi) - \alpha(0) = 2\pi n \ . \tag{2.8}$$

Given (2.7) the asymptotic behaviour of the gauge field follows from the vanishing of the second term

$$\lim_{\rho\to\infty} D_{\mu}\Phi \simeq 0 \quad \Longrightarrow \quad A_{\mu} = -\frac{1}{e}\partial_{\mu}\alpha \ . \tag{2.9}$$

Apparently A_μ approaches a pure gauge, which implies the integrability of the first term as well.

Let us calculate the magnetic flux Φ in the x_3 direction for a configuration with the asymptotic behaviour specified above. Using Stokes theorem one obtains

$$\Phi = \int \vec{B}.d\vec{s} = \oint \vec{A}.d\vec{\ell} = \frac{1}{e}\left[\alpha(2\pi)-\alpha(o)\right] = \frac{2\pi n}{e} . \quad (2.11)$$

An important result. If there exists a time and x_3 independent solution with finite energy per unit length (length is of course measured in the x_3 direction) then this solution has a magnetic flux which is quantized in units $\frac{2\pi}{e}$. Indeed, in the type II superconductor for which the Lagrangian (2.1) provides a good description, these Abrikosov (or Nielsen-Olesen) flux tubes[3] do occur. Applying a magnetic field over a superconductor one sees that the flux gets squeezed into narrow tubes each containing a single unit of flux.

3. TOPOLOGY OF THE NIELSEN-OLESEN FLUX TUBE

The remarkable fact that the total flux is completely independent of the details of the solution for finite values of ρ (assuming it exists) suggests that it may have a topological interpretation. We give the argument in some detail because we will apply it in more general cases later on. Observe that $\Phi(\rho\to\infty)$ may be considered as a mapping from the boundary of space (i.e. the circle S^1_x) to the vacuum manifold (S^1_Φ)

$$\Phi(\rho\to\infty) \; : \; S^1_x \to S^1_\Phi \quad (3.1)$$

It assigns to every ϕ a value $\alpha(\phi)$. The condition (2.8) implies that if we go around S^1_x once then α winds around S^1_Φ exactly n times. This winding n is a topological invariant. All the single valued maps Φ of $S^1_x \to S^1_\Phi$ fall into distinct homotopy classes $\{\;\}_n$ labeled by the winding number n. It is impossible to continuously deform a map of a given class labeled by n into one from a class $n' \neq n$. The topological charge Q defined as

$$Q = \frac{e}{2\pi}\oint \vec{A}.d\vec{\ell} = n \quad (3.2)$$

is just the winding number (or Brouwer degree) of the map. The boundary condition allows for a compactification of R^2 to S^2, in which case (3.2) is the first Chern class of the bundle(s) with connection A_μ.

A basic theorem of homotopy theory states that the classes $\{\ \}_n$ themselves form a (discrete) group, the first homotopy or fundamental group denoted by π_1. It is easy to see that if

$$\Phi_1 \in \{\ \}_{n_1} \quad , \quad \Phi_2 \in \{\ \}_{n_2} , \tag{3.3}$$

and

$$\hat{\Phi} = \Phi_1 \Phi_2 \quad , \tag{3.4}$$

then

$$\hat{\Phi} \in \{\ \}_{n_1+n_2} \quad . \tag{3.5}$$

Hence

$$\pi_1(U_1) = Z , \tag{3.6}$$

where Z is the group of integers under addition.
The fact that n is invariant under continuous variations of the fields implies (at least on a classical level) the existence of a globally conserved "quantum" number , In other words, the space of finite energy solutions corresponding to the theory (2.1) breaks up into disconnected sectors labeled by the charge n. A (continuous) time evolution of the system does not allow one to go from n to n' ≠ n. It is important to note that this charge comes about in an entirely different manner as the charges we discussed in the previous lecture. Those "Noether charges" generate a continuous symmetry of the Lagrangian density. These "topological charges" are a direct consequence of the topology of the vacuum and space time manifolds.

It is clear that from topological considerations alone nothing can be concluded about the existence of finite energy solutions. Though it is a strong hint of where to expect them, an existence proof or actual construction of solutions depends on the details of the equations.

4. TOWARDS NONSINGULAR SOLUTIONS

Let us consider the model in more detail. It is convenient to define the dimensionless variables

$$A^{\pm} = \frac{1}{2f}(A_1 \mp iA_2) \ ,$$
$$\phi = \Phi/f \ , \tag{4.1}$$

and

$$z^{\pm} = ef(x_1 \pm ix_2) \ . \tag{4.2}$$

There is a special choice of the mass ratio m_A/m_χ (Landau parameter)

$$m_A/m_\chi = e/\sqrt{\lambda} = \sqrt{2} \ , \tag{4.3}$$

for which it is possible to derive a lower bound for the energy per unit length E. In dimensionless units it can be casted in the form

$$\varepsilon = \frac{e}{\pi f} E = \frac{1}{\pi} \int dz^+ dz^- \left\{ (\partial_- - iA_-)\phi\,(\partial_+ + iA_+)\phi^\dagger + \right.$$
$$\left. + \left[\tfrac{1}{4}(\phi^2 - 1) - i(\partial_- A_+ - \partial_+ A_-)\right]^2 \right\} + n \ . \tag{4.4}$$

Because of the positivity of the first two terms we saturate the lower bound $\varepsilon = n$ $(n > o)$ if and only if both terms vanish identically:

$$(\partial_- - iA_-)\phi = o \ , \tag{4.5a}$$

$$(\partial_+ + iA_+)\phi^\dagger = o \ , \tag{4.5b}$$

$$\partial_+ A_- - \partial_- A_+ + \frac{i}{4}(\phi^\dagger\phi - 1) = o \ . \tag{4.5c}$$

Solutions of these first order (!) equations are also solutions to the second order field equations because they correspond by construction to extrema of the two dimensional action. Note that there is apparently no interaction energy between the fluxtubes. Indeed, it is the limit where fluxtubes do not interact (the transition point between type I and type II superconductors). Further simplification occurs if one chooses a suitable gauge condition, for example (without loss of generality) one may impose the Coulomb gauge

$$\vec{\nabla}.\vec{A} = \partial_+A_- + \partial_-A_+ = o \; , \tag{4.6}$$

which is solved by

$$A_\pm = \pm i \; \partial_\pm\psi \; . \tag{4.7}$$

Upon substitution into equations (4.5a-b) these reduce to the Cauchy-Riemann conditions for the function

$$g = e^{-\psi}\Phi \; . \tag{4.8}$$

With the definition of a new field χ where

$$\psi = \tfrac{1}{2}(\chi - \ell n\, g - \ell n\, g^*) \; , \tag{4.9}$$

we obtain from (4.5c) the final equation for χ

$$\partial_+\partial_-\chi = e^\chi - 1 \; . \tag{4.10}$$

The boundary conditions are:

$$\begin{aligned} &\chi \to o \quad (|z| \to \infty) \; , \\ &\chi \to \tfrac{1}{2}\ell n \; |z-z_i| \qquad (z \to z_i) \; . \end{aligned} \tag{4.11}$$

Strictly speaking some $\delta(z-z_i)$ type source terms should be included in the right hand side of (4.10) because of the logarithmic behaviour of χ near $z = z_i$. These do not lead to singularities in gauge invariant hence physically interesting quantities.

What is known about the system (4.10-11)? E. Weinberg[4)] using some variant of the Atigah-Singer index theorem has shown that the solution with $Q = n$ has $N = 2n$ parameters. These correspond presumably to the coordinates of n independent fluxtubes with $Q = 1$. Taubes[5)] has given an existence proof for a finite energy static solution with n separated fluxtubes. Exact solutions are not known explicitly, even not for the n = 1 case.

5. CONFINEMENT OF MAGNETIC CHARGE

In the previous lecture we alluded to the for experimentalists rather traumatic situation that a quark cannot be observed as a free particle. This problem of quark confinement has some resemblance with that of a bar magnet. Upon breaking one bar magnet, one obtains two bar magnets etc. It is impossible to isolate a magnetic north (or south) pole. Nambu pointed out that in a superconductor magnetic monopoles (if they existed) would be permanently confined. The magnetic flux emanating from the pole would be squeezed into a Nielsen-Olesen fluxtube. The energy of the tube, ℓE, (4.4) is proportional to its length ℓ and hence leads to a linear potential between a pole-antipole pair.

It is instructive to discuss this phenomenon in more detail. It necessitates the introduction of the magnetic monopole â la Dirac. He gave a prescription for introducing the magnetic monopole into electrodynamics but still maintaining the vector potential A_μ. The usual definition of the magnetic field

$$\vec{B} = \vec{\nabla} \times \vec{A} \tag{5.1}$$

necessarily implies

$$\vec{\nabla} . \vec{B} = 0 \tag{5.2}$$

and has therefore to be modified in order to have monopoles. Dirac considered a gauge potential of the form

$$\vec{A} = \frac{g}{4\pi} \vec{A}^D = \frac{g}{4\pi} \frac{\hat{r} \wedge \hat{n}}{r(1-\hat{r}.\hat{n})} \; . \tag{5.3}$$

Its curl yields a radial magnetic field of strength g emanating from the endpoint at r = o of an infinitely thin straight solenoid in the direction $\hat{n}$. One now defines

$$F_{\mu\nu} = \partial_\mu A_\nu - \partial_\nu A_\mu - G_{\mu\nu} \tag{5.4}$$

where $G_{\mu\nu}$ is the singular Dirac string, carefully chosen as to cancel the solenoidal field along the $\hat{n}$ direction, so that only the monopole field survives.

We exploit this idea to construct an approximate solution corresponding to a spacially separated monopole - antimonopole pair in the Abelian Higgs model. Incorporating (5.4), the field equations take the form

$$\Box A_\nu = \partial_\mu G_{\mu\nu} - ie\,\Phi^\dagger D_\nu \Phi \tag{5.5}$$

in the gauge $\partial_\mu A_\mu = 0$. Let us take the situation where a monopole is located at $\vec{x} = (0,0,a)$ and an antimonopole at $\vec{x} = (0,0,-a)$. For the corresponding string we choose

$$G_{\mu\nu} = -\varepsilon_{0\mu\nu 3}\, g\, \delta(x)\, \delta(y) \left[\theta(a-z) - \theta(-a-z)\right] \tag{5.6}$$

where θ is the stepfunction. Not to close to the singularities of $G_{\mu\nu}$ we can assume $\Phi = f$. This generates the mass term for A_μ in equation (5.5):

$$\left(\vec{\nabla}^2 - m_A^2\right)\vec{A} = \vec{j} \tag{5.7}$$

with $m_A = ef$ and $j_\nu = \partial_\mu G_{\mu\nu}$. The solution is given by

$$\vec{A} = \int d^3x'\; K(\vec{x}-\vec{x}')\; \vec{j}(\vec{x}') \tag{5.8}$$

where the Green's function

$$K(\vec{x}-\vec{x}') = \frac{e^{-m_A|\vec{x}-\vec{x}'|}}{|\vec{x}-\vec{x}'|} \tag{5.9}$$

satisfies the equation

$$\left(\vec{\nabla}^2 - m_A^2\right) K(\vec{x}-\vec{x}') = \delta(\vec{x}) \tag{5.10}$$

as well as the appropriate boundary conditions. Using (5.6) and (5.7) we partially integrate (5.8) to obtain

$$\vec{A} = -g \int_{-a}^{a} dz'\; \hat{z} \times \left[\vec{\nabla} K(\vec{x}-\vec{x}')\right]_{\vec{x}' = z'\hat{z}} \;. \tag{5.11}$$

To study the behaviour of the magnetic field $\vec{B} = \vec{\nabla} \times \vec{A}$ we exploit the identity

$$\vec{\nabla} \times (\hat{z} \times \vec{\nabla}) K = \hat{z}\nabla^2 K - (\hat{z}.\vec{\nabla})\vec{\nabla} K = \hat{z}\, m_A^2 K - \partial_z(\vec{\nabla} K) \;, \tag{5.12}$$

so that

$$\vec{B} = -g \left[\hat{z}\, m_V^2 \int_{-a}^{a} dz'\; K(\vec{x}-\vec{x}') + \vec{\nabla}K(\vec{x}-\vec{a}) - \vec{\nabla}K(\vec{x}+\vec{a}) \right] , \qquad (5.13)$$

with $\vec{x}' = z'\hat{z}$.

In the neighbourhood of each pole ($\vec{x} \to \pm\vec{a}$) the leading contribution comes from the gradient terms yielding the usual radial magnetic field

$$\vec{B} = \mp g\; (\vec{x}\pm\vec{a})/|\vec{x}\pm\vec{a}|^3 \qquad (5.14)$$

up to a radius $|\vec{x}\pm\vec{a}| \simeq 1/m_A$, where the field lines bend over into a fluxtube which connects the monopoles. In a region close to the core of the tube but far from the poles ($m_A|\vec{x}\pm\vec{a}| \gg 1$) we can approximate the magnetic field in cylindrical coordinates

$$\vec{B} = -g\,\hat{z}\, m_A^2 \int_{-\infty}^{\infty} dz\; \frac{\exp(-m_A\sqrt{\rho^2+z^2})}{\sqrt{\rho^2+z^2}}$$

$$\equiv -2g\,\hat{z}\, m_A^2\; K_o(m_A\rho) \;, \qquad (5.15)$$

where $K_o(z)$ is a modified spherical Bessel function with asymptotic behaviour

$$K_o(z \to \infty) \simeq \sqrt{\frac{\pi}{2z}}\; e^{-z} \left\{ 1+O(\tfrac{1}{z}) \right\} . \qquad (5.16)$$

The width of the fluxtube is thus of order $1/m_A$ and its mass per unit length proportional to $g\, m_A^2$. To obtain a complete picture of the profile of the tube one has to solve the nonlinear system (4.10). To also have non singular monopoles at the endpoints it is necessary to go to a nonabelian group G as will become clear later on.

6. THE TOPOLOGY OF GAUGE FIELDS ON R^n

We consider a (spontaneously broken) gauge theory with an arbitrary compact gauge group G,

$$\mathcal{L} = -\tfrac{1}{4}\left(F^a_{\mu\nu}\right)^2 - \tfrac{1}{2}\left(D_\mu \Phi\right)^2 - V(\Phi) \tag{6.1}$$

and pose the question whether there are topological excitations in a Euclidean space time $M = R^d$. As before we demand that the terms of (6.1) vanish separatedly on the boundary ∂M. In particular

$$\left.\frac{\partial V}{\partial \Phi}\right|_{x \in \partial M} = 0 \ . \tag{6.2}$$

If the minimum of the potential is achieved for some $\Phi = \Phi_0$ then it is minimized for all $\Phi = g\Phi_0$ ($\forall_g \in G$). The residual symmetry group H (i.e. the stability group of Φ_0) is defined as

$$H = \left\{h \ ; \ h \in G \ , \ h\Phi_0 = \Phi_0\right\} \ . \tag{6.3}$$

In other words the generators of H annihilate Φ_0. We may parametrize $g \in G$ as

$$g = kh \quad , \quad \begin{matrix} h \in H \\ k \in G/H \end{matrix} \ , \tag{6.4}$$

and obtain the vacuum manifold

$$V = \left\{g\Phi_0\right\} = \left\{k\Phi_0\right\} \simeq G/H \ . \tag{6.5}$$

It is isomorphic to the coset space G/H. Bearing in mind that Φ is not necessarily constant on ∂M

$$\Phi(\partial M) = k(\partial M)\Phi_0 \tag{6.6}$$

we are led to consider the maps

$$k(\partial M) \ : \ \partial M \to G/H \ . \tag{6.7}$$

Thinking of ∂M as a hypersphere S^{d-1}_x the topology of k is determined by the nth homotopy group ($n=d-1$)

$$\pi_n(G/H) \ . \tag{6.8}$$

If this group for some choice of M, G and H turns out to be nontrivial, it implies the existence of some topological charge. It does however not imply the existence of finite energy (action) solutions.

The behaviour of the gauge fields are determined by the condition of a vanishing covariant derivative on the boundary

$$D_\mu \Phi(\partial M) = \left[\partial_\mu k\, k^{-1} + iA_\mu\right] k\Phi_o = o \ . \tag{6.9}$$

If we split A_μ as

$$A_\mu = A_\mu^{\parallel} + A_\mu^{\perp} \tag{6.10}$$

where $A_\mu^{\parallel}$ contains the generators of H and $A_\mu^{\perp}$ the generators of G/H, then (6.9) determines that asymptotically

$$A_\mu^{\perp} = -\, i(\partial_\mu k)k^{-1} \tag{6.11}$$

but leaves $A_\mu^{\parallel}$ undetermined since

$$A_\mu^{\parallel}\Phi_o = o \ . \tag{6.12}$$

This behaviour is to be expected because the Higgs mechanism generates a mass for $A_\mu^{\perp}$ but not for $A_\mu^{\parallel}$. The asymptotic behaviour of $A_\mu^{\parallel}$ is determined by the integrability of the term involving $F^a_{\mu\nu}$. In three dimensions it may have nontrivial long range components giving rise to an asymptotic radial (electric) magnetic field. In four dimensions also $A_\mu^{\parallel}$ should approach a pure gauge and the topological classification of instantons is not affected by the residual symmetry group H. We summarize the results with some typical examples in the table.

	M	∂M	V		Top. charge
Fluxtube	R^2	S^1	U_1	$\pi_1(S^1) = Z$	Magn. Flux
Monopole	R^3	S^2	SO_3/U_1	$\pi_2(S^2) = Z$	Magn. Charge
Instanton	R^4	S^3	SU_N	$\pi_3(S^3) = Z$	Pont. Index

The analysis we just presented has the advantage of being simple but has its drawbacks for example as it comes to discussing the topology of gauge fields in a manifold M without boundary like spheres and tori. In these cases it is necessary to construct a fibre bundle over the base manifold. The bundle topology is characterized by the Chern classes. In many cases properties of the solutions on R^n can be related to those of solutions on some compactification. For example if A_μ approaches a pure gauge asymptotically in R^n one may as well consider S^n. On compact manifolds powerful index theorems can be used to for example relate the dimensionality of the solution space and the value of the topological charges. In some cases there is a direct mapping of solutions. From a solution on S^4 one obtains a solution on R^4 by stereographic projection. The study of Yang-Mills fields on the hypertorus T^4 initiated by 't Hooft[6] has revealed important information on the phase structure of nonabelian gauge theories. Another example is the problem of gauge fields at finite (real) temperature which involves the Yang-Mills equations on $R^3 \times S^1$ [7].

7. THE EXACT SEQUENCE

A useful tool for the determination of $\pi_n(G/H)$ is the homotopy sequence. It is a sequence of homomorphisms between the various homotopy groups,

$$\ldots \rightarrow \pi_n(H) \rightarrow \pi_n(G) \rightarrow \pi_n(G/H) \rightarrow \pi_{n-1}(H) \rightarrow \pi_{n-1}(G) \rightarrow \ldots \qquad (7.1)$$

with the property that for two successive mappings in the sequence the image (Im) of the first map equals the kernel (Ker) of the second.

For example in the case of monopoles it is convenient to re-express $\pi_2(G/H)$. It can be shown that

$$\pi_2(G) \simeq \pi_2(S^3) = 0 \qquad (7.2)$$

for all compact Lie groups. It then follows from

$$\mathrm{Im}\left[\pi_2(G) \rightarrow \pi_2(G/H)\right] \simeq \mathrm{Ker}\left[\pi_2(G/H) \rightarrow \pi_1(H)\right] \qquad (7.3)$$

that the map $\pi_2(G/H) \rightarrow \pi_1(H)$ is one to one. This implies that

$$\mathrm{Im}\left[\pi_2(G/H) \rightarrow \pi_1(H)\right] \simeq \pi_2(G/H)\ , \qquad (7.4)$$

or in other words:

$$\pi_2(G/H) \simeq \mathrm{Ker}\left[\pi_1(H) \to \pi_1(G)\right] . \tag{7.5}$$

A further simplification occurs if G is simply connected (i.e. $\pi_1(G) = o$) in which case one obtains the isomorphism

$$\pi_2(G/H) = \pi_1(H) . \tag{7.6}$$

The monopole charges are completely determined by the connectivity properties of H.

8. Z_N MAGNETIC FLUX

Instead of studying the space G/H we may look at $\bar{G}/\bar{H}$, where $\bar{G}$ is the simply connected covering group of G and $\bar{H}$ the lift of H into $\bar{G}$. By an argument similar to the one presented in the previous section one can prove the isomorphism,

$$\pi_1(G/H) \simeq \pi_o(\bar{H}) \simeq \bar{H} \tag{8.1}$$

assuming in the last equality that $\bar{H}$ is actually discrete. Let us consider some examples of spontaneously broken theories in R^2. We assume $G = SU_2$.

a) Consider the case where a scalar doublet Φ acquires a nonvanishing expectation value. $<\Phi>$ breaks the group completely hence $G/H \simeq G \simeq SU_2$. The group manifold of SU_2 is a three sphere S^3, so that

$$\pi_1(SU_2) \simeq \pi_1(S^3) = o . \tag{8.2}$$

This simple computation shows that the Glashow-Weinberg-Salam model of the weak and electromagnetic interactions does not have topological fluxtubes.

b) The case where Φ is taken to be a triplet is not much better. The residual symmetry group H is the U_1 group of rotations around the axis $<\Phi>$. From (8.1) it follows that

$$\pi_1(\bar{G}/\bar{H}) = \pi_o(U_1) = o \tag{8.3}$$

c) Now introduce two triplets ϕ and χ both with non vanishing expectation values (making some arbitrary angle). Then $\bar{H} = Z_2$, the group of integers modulo 2. The nontrivial element correspond to a rotation of 2π, which leaves the triplets invariant, but corresponds to the center element (-1) of SU_2 because a doublet changes sign under a 2π rotation. Consequently,

$$\pi_1(SU_2/Z_2) = Z_2 \ . \tag{8.4}$$

Apparently there is a topological flux which is conserved modulo 2. Two units of flux are topologically equivalent to no flux at all. Physically it means that whereas a single unit of flux is really forced into a tube, two units of flux can be screened or annihilated by the medium. This remarkable property of the nonabelian superconductor was first noticed by Mandelstam.

A similar situation arises if one breaks SU_N down to its center Z_N. The fact that

$$\pi_1(SU_N/Z_N) = Z_N \tag{8.5}$$

implies the existence of flux which is conserved modulo N. The reader may recall the enigma of quark confinement: The property that quarks are only observed in the color singlet combinations quark-antiquark (mesons) and three quarks (baryons). If quarks would carry Z_3 magnetic charge and the color group SU_3 would be broken down to Z_3, then these magnetic quarks would be permanently confined. However in QCD we want SU_3 to be unbroken and the quarks to carry electric charge. The groundstate should therefore rather have the properties of a magnetic superconductor.

9. NONABELIAN FLUX

The homotopy groups π_n are discrete abelian groups if $n > 2$. For π_1 there is the odd possibility of being nonabelian. From (8.1) one directly learns that this is the case if $\bar{H}$ is nonabelian. As the homotopy group tells us how to "add" charges it is interesting to ask what the physical implications are for the fluxtubes associated with the elements of $\bar{H}$ [8].

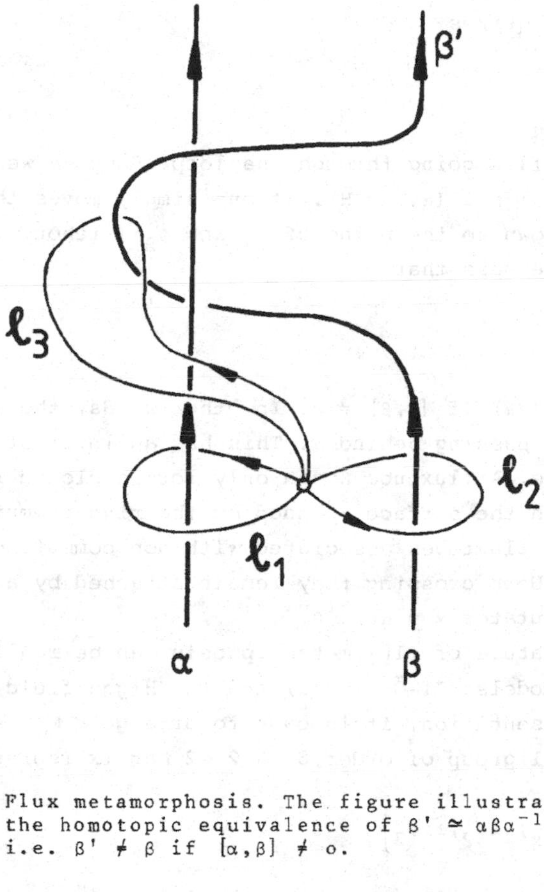

Flux metamorphosis. The figure illustrates the homotopic equivalence of $\beta' \simeq \alpha\beta\alpha^{-1}$, i.e. $\beta' \neq \beta$ if $[\alpha,\beta] \neq o$.

In the figure we have drawn three closed paths ℓ_i (i=1,2,3) in space, each encircling one of two different fluxtubes. We parametrize the loops ℓ_i by parameters ϕ_i ($o \leq \phi_i < 2\pi$) with $\ell_i = o$ corresponding to the interaction of the loops. Consider the Higgs field ϕ. Far enough from the core of the fluxtubes, in particular on the curves ℓ_i, the covariant derivative $D_\mu \Phi$ vanishes. A formal solution can be written as

$$\Phi(\phi_i) = U_i(\phi_i)\ \Phi(o) \ , \tag{9.1}$$

with the path ordered exponential

$$U_i(\phi_i) = P\, e^{ie \int_o^{\phi_i} A_{\phi_i} d\phi_i} \ . \tag{9.2}$$

In other words Φ is parallel transported along the loops.
Single valuedness requires

$$U_i(2\pi) \in \bar{H} \tag{9.3}$$

i.e. it labels the flux going through the loop. Suppose we start with $U_1(2\pi) = \beta$ and $U_2(2\pi) = \alpha$ $(\alpha, \beta \in \bar{H})$. If one simply moves the third loop continuously down to the plane of ℓ_1 and ℓ_2 (without crossing the tubes of course) one sees that

$$U_3(2\pi) = \alpha\beta\alpha^{-1} \ . \tag{9.4}$$

Clearly $U_3(2\pi) \neq U_1(2\pi)$ if $[\alpha, \beta] \neq o$. In other words, the flux β has changed identify by passing behind α. This has an interesting physical consequence. A fluxtube β can only form a closed ring if the total flux α through the surface spanned by the ring commutes with β. Stated differently, fluxtubes associated with non commuting elements of $\bar{H}$ cannot cross. Upon crossing they remain attached by a third tube labeled by the commutator $\gamma = \alpha\beta\alpha^{-1}\beta^{-1}$.

This exotic feature of flux metamorphosis can be realized in relatively simple models. If $\bar{G} = SU(2)$ and the Higgs field Φ forms a 5 dimensional representation, it is easy to arrange $\langle \Phi \rangle$ such that $\bar{H} = \bar{D}_2$, the dihedral group of order 8. A 2×2 matrix representation is

$$\bar{D}_2 = \left\{ 1, -1, \pm i\sigma_x, \pm i\sigma_2, \pm i\sigma_3 \right\} \tag{9.5}$$

where the σ_i are the traceless hermitean Pauli matrices.

10. MAGNETIC MONOPOLES

We have shown in section 7 how the monopole charges are associated with the elements of

$$\pi_2(G/H) = \pi_2(\bar{G}/\bar{H}) = \pi_1(\bar{H}) \tag{10.1}$$

in particular by connectivity properties of $\bar{H}$, the lift of H to the simply connected covering group $\bar{G}$ of G. Given the group G the maximal number of conserved magnetic charges occurs if we break G down to its maximal torus, that is to the maximal abelian subgroup contained in G,

$$\bar{H} = \left[\otimes U_1\right]^{\ell} \tag{10.2}$$

with ℓ = rank G. The fundamental group of $\bar{H}$ is

$$\pi_1(\bar{H}) = \left[\otimes Z\right]^{\ell} , \tag{10.3}$$

implying that the charge is labeled by ℓ-integers. We will return to this question in sections 14 and 15 in all its generality, but limit ourselves now to the observation that the simplest model is obtained with $\bar{G} = SU_2$ and $\bar{H} = U_1$. This is indeed the prototype model where the existence of a nonsingular monopole solution was discovered by 't Hooft and Polyakov (1974)[9]. It is a remarkable state of affairs. One starts with a model which contains only electrically charged fundamental fields which because of the nonabelian nature of the underlying theory can form boundstates which are magnetic monopoles. The monopoles are not put in by hand, they are an unavoidable ingredient of any theory where the U_1 group of the electromagnetic interactions is realized as a subgroup of some simple unifying group G. This is actually believed to be the case, as it accounts for the observed quantization of electric charge. Possibly the only reason that monopoles have not yet been observed is that they would be extremely massive.

We note that there are other possibilities for $\bar{H}$ than just a product of U_1 factors. The simplest example is a theory with $\bar{G} = SU_3$ broken down to $\bar{H} = SO_3$ in which case

$$\pi_1\left[SO_3\right] = Z_2$$

i.e. the charge is conserved modulo two.

11. THE 't HOOFT-POLYAKOV MONOPOLE

Consider a theory with $G = SO_3$ and a scalar field in the adjoint (triplet) representation. Assuming a potential of the form

$$V(\phi) = \frac{\lambda}{4}\,(\vec{\Phi}.\vec{\Phi} - f^2)^2 \; . \tag{11.1}$$

It is minimized for any $\langle\vec{\Phi}\rangle = \vec{f}$ where $\vec{f}$ is some vector of length f. The mass terms for the gauge fields are easily obtained from

$$\left(e\vec{A}_\mu \times \vec{f}\right)^2 = e^2 f^2 \left[\vec{A}^2 - \left(\vec{A}_\mu . \hat{f}\right)^2\right] = e^2 f^2 \left(A_\mu^\perp\right)^2 \; . \tag{11.2}$$

The two components orthogonal to $\langle\vec{\Phi}\rangle$ obtain a mass ef. The component along $\langle\vec{\Phi}\rangle$, corresponding to the generator of the residual symmetry group $H = U_1$ of rotations around $\langle\vec{\Phi}\rangle$, remains massless. At low energies ($E \ll ef$) this is just the abelian Maxwell theory, but as we will see enriched with magnetic monopoles.

The vacuum manifold G/H is just the two sphere S^2_Φ, the orbit of $\langle\vec{\Phi}\rangle$ under the group SO_3. The mappings $\langle\vec{\Phi}\rangle : S^2_x \to S^2_\Phi$ fall into homotopy classes labeled by an integer as

$$\pi_2(G/H) = \pi_2(S^2) = Z \; . \tag{11.3}$$

For example a map of degree n = 1 would be

$$\vec{\Phi} = f\,\frac{\vec{x}}{x} \; . \tag{11.4}$$

The gauge potential obtained from $D_\mu\vec{\Phi} = o$ is of the form

$$eA_i = \frac{1}{f^2}(\partial_i\vec{\Phi}) \times \vec{\Phi} + a\vec{\Phi} \tag{11.5}$$

yielding (a = o),

$$eA_i^a = \varepsilon_{aib}\, x_b/x^2 \; . \tag{11.6}$$

Note the way real space and internal space indices are intertwined in the formulae (11.4) and (11.6).

We have set $A_o = o$ and assumed time independence, anticipating a static purely magnetic solution. Static solutions with $A_o \neq o$ exist as well. They correspond to "dyons", carrying both electric and magnetic charge[10].

It is remarkable that the behaviour (11.4) and (11.6) of the fields $\vec{\Phi}$ and $\vec{A}_\mu$ do not involve a Dirac type string. Indeed, if the U_1 gauge group is embedded in a nonabelian group one can avoid the string in certain cases. This is basically the content of formula (10.1). If $H = U_1$, then all allowed strings (monopoles) are labeled by elements of $\pi_1(H)$. Potentially nonsingular monopoles for which the string can be avoided (removed) correspond to elements of $\pi_2(G/H) = \pi_1(\bar{H})$. The elements of $\pi_1(H)$ which are not contained in its subgroup $\pi_1(\bar{H})$ correspond to Dirac monopoles for which the string cannot be avoided.

Before entering in a discussion of the nonsingular solutions we elaborate some more on the topological meaning of the magnetic charge.

12. TOPOLOGY OF THE CHARGE

In order to calculate the magnetic charge explicitly one first has to give a gauge (SO_3) invariant definition of the abelian field strength $f_{\mu\nu}$ associated with the unbroken U_1 subgroup. 't Hooft suggested[9] the following expression ($\hat{\phi}=\vec{\Phi}/|\vec{\Phi}|$),

$$f_{\mu\nu} = \hat{\Phi}.\vec{F}_{\mu\nu} - \frac{1}{e}\frac{\hat{\Phi}}{\Phi^2}.(D_\mu\vec{\Phi}\times D_\nu\vec{\Phi}) \ . \tag{12.1}$$

The presence of the second term can be understood as follows. Suppose $\hat{\phi}$ is constant e.g. $\hat{\Phi}^a = \delta^a_3$ then $f_{\mu\nu}$ is just the curl of A^3_μ,

$$f_{\mu\nu} = \partial_\mu A^3_\nu - \partial_\nu A^3_\mu \ . \tag{12.2}$$

The first term of (12.1) includes also the piece $e\vec{A}_\mu\times\vec{A}_\nu.\hat{\Phi}$. The second term is the unique gauge invariant term which cancels it. To calculate the total magnetic charge one only needs the asymptotic form of f_{ij}. Since $D_i\vec{\Phi} = o$ one has (using 11.5)

$$f_{ij} = \hat{\Phi}.\vec{F}_{ij} = \frac{1}{e}\,\hat{\Phi}.(\partial_i\hat{\Phi}\times\partial_j\hat{\Phi}) \ . \tag{12.3}$$

Again this expression is completely determined by $\hat{\Phi}$ and hence independent of detailed properties of the solution. The magnetic field $B_i = \frac{1}{2}\,\varepsilon_{ijk}f_{jk}$ can be integrated over a closed surface S^2_x at infinity to yield the charge

$$g = \frac{1}{e}\int_{S_x} \tfrac{1}{2}\,\varepsilon_{ijk}\,\hat{\Phi}.(\partial_j\hat{\Phi}\times\partial_k\hat{\Phi})\ d\sigma^i \ . \tag{12.4}$$

Here $d\sigma^i$ denotes the surface element orthogonal to $\hat{x}_i$,

$$\varepsilon_{ijk}\, d\sigma^i = \frac{\partial(x^j,x^k)}{\partial(\xi_1,\xi_2)}\, d\xi_1\, d\xi_2 \tag{12.5}$$

with

$$\frac{\partial(x^j,x^k)}{\partial(\xi_1,\xi_2)} = \frac{\partial x^j}{\partial \xi_1}\frac{\partial x^k}{\partial \xi_2^2} - \frac{\partial x^j}{\partial \xi^2}\frac{\partial x^k}{\partial \xi^1} \tag{12.6}$$

i.e. the Jacobian of the transformation from x^i to the parameters ξ_1 and ξ_2 of the surface. We rewrite (12.4) now as

$$g=\frac{1}{e}\int_S \tfrac{1}{2}\, \varepsilon^{abc}\, \hat{\phi}^a \frac{\partial(\hat{\phi}^b\hat{\phi}^c)}{\partial(\xi_1\xi_2)}\, d\xi_1\, d\xi_2$$

$$= \frac{1}{e}\, n \int_{S_\phi} \hat{\phi}^a\, d\sigma^a_\phi = \frac{4\pi n}{e} . \tag{12.7}$$

This computation shows explicitly that in a situation where we have no string the magnetic charge is completely determined by the topology of the Higgs field $\hat{\phi}$. Note that the charge (12.7) has twice the value of the minimal allowed U_1 charge $2\pi/e$. It reflects the fact that in this case $\pi_1(\bar{H})$ only contains the even elements of $\pi_1(H)$.

13. THE BOGOMOLNY-PRASAD-SOMMERFIELD LIMIT

The expression for the energy (mass) of the monopole

$$E = \int d^3x \left[\frac{\vec{B}_i^2}{2} + \tfrac{1}{2}(\vec{D}_i\Phi)^2 + V(\Phi) \right] \tag{13.1}$$

can be rewritten in the suggestive form

$$E = \int d^3x \left\{ \tfrac{1}{2}\left[\vec{B}_i \pm D_i\vec{\Phi}\right]^2 \mp \vec{B}_i . D_i\vec{\Phi} + V(\Phi) \right\} . \tag{13.2}$$

The second term is easily shown to be proportional to the magnetic charge

$$\int d^3x\, \vec{B}_i . D_i\vec{\Phi} = \lim_{|r|\to\infty} \left[r^2 \int d^2\Omega\, \vec{B}_r . \vec{\Phi} \right]$$

$$- \int \vec{\Phi} . D_i\vec{B}_i\, d^3x = gf \tag{13.3}$$

because of the Bianchi identity $D_i\vec{B}_i \equiv o$. The construction of solutions is now tremendously simplified if one takes the limit $\lambda \to o$ in $V(\Phi)$. It is the limit where the mass $m_\Phi = \sqrt{\lambda}\, f$ of the surviving scalar particle vanishes. Since also $V(\Phi) \to o$ the mass of a monopole with charge g is bounded

$$E \geqslant |g| f . \tag{13.4}$$

The lower bound is in fact saturated for any solution of the first order Bogomolny equations[11)]

$$\vec{B}_i = \pm D_i\vec{\Phi} . \tag{13.5}$$

These are much simpler than the full second order field equations. Note that in the B.P.S. limit there is no interaction energy associated with monopoles whose charges have the same sign. The physical reason is that the repulsive Coulomb force between equal sign charges due to the long range (massless) component of the gauge potential is cancelled by an attractive force due to the massless scalar particle.

We also learn that the mass of the monopole equals

$$E = |g| f = \frac{4\pi n}{e^2} m_A \tag{13.6}$$

i.e. typically several orders of magnitude larger than the mass which is generated for the gauge particles. As we mentioned before in the Glashow-Weinberg-Salam model the vacuum manifold $G/H \simeq S^3$, so that $\pi_2(G/H) = o$. Consequently this model has no monopoles. In grand unified theories (GUTs) one usually starts with a simple group G which breaks down to $H = U_1 \times K$ with K some semi-simple group without U_1 factor. The scale rate at which this breaking has to occur is typically $m_A \simeq 10^{14} - 10^{15}$ GeV. So in such a scheme there are monopoles with a mass of $m \simeq 10^{16}$ GeV which in "household" units corresponds to 10^{-7} grams!

What is known about the solutions of the Bogomolny equations (13.5) for the simple SO_3 model. For a long time only the exact spherically symmetric n = 1 solution found by Prasad and Sommerfield[12] (1975) was known. Writing the ansatz for the fields as

$$eA^a_{\mu} = \varepsilon_{aib} \frac{x_b}{x^2} (1-xH) \ ,$$

$$\phi^a = \frac{x^a}{x} (1+xQ) \ ,$$

where H and Q are only functions of $x = |x|$ the Bogomolny equations take the simple form

$$H' = HQ \ ,$$

$$Q' = H^2 \ ,$$

with the solution (f = 1)

$$H = 1/\sinh x$$
$$Q = - \coth x \ .$$

E. Weinberg[13] showed that the general solution of charge $g = \frac{4\pi n}{e}$ should contain N = 4n - 1 parameters. Subsequently Taubes[14] proved the existence of static finite energy solutions corresponding to n specially separated monopoles. Recently exact solutions describing spatially separated multimonopole solutions have been constructed for the SO_3 model. These are based on an adaption of Penrose's twistor formalism by Ward and Atiyah[15]. This construction is described in detail in the lectures of R.S. Ward and E. Corrigan.

14. MAGNETIC CHARGE QUANTISATION

We have stressed the intimate connection between the magnetic charge and the nontrivial topology of the Higgs field Φ. One may imagine a gauge transformation such that Φ is constant. Exactly because in this new gauge Φ would have a trivial topology the gauge transformation has to be singular. As one may expect it reproduces the (in)famous Dirac string in some of the vector potentials belonging to H. To obtain the maximal number of topologically conserved components of the magnetic charge we break G down to the maximal torus

$$G \supset H \approx (\otimes U_1)^{\ell} \qquad (14.1)$$

where ℓ = rank G. The torus is generated by the Cartan subalgebra C of G. The quantisation of magnetic charge is then expressed by the condition (very similar to Dirac's original one)

$$e^{ieg^a H^a} = 1 \quad , \quad H^a \in C \qquad (14.2)$$

where e is the coupling constant and g^a $(a = 1,\ldots,e)$ are the ℓ components of the magnetic charge.

The allowed values of eg^a in (14.2) depend of course on the representation of H^a. This allows the distinction between the 't Hooft-Polyakov monopoles (for which the string can be removed) and Dirac monopoles (which necessarily involve a string). All monopole charges g^a have to satisfy (14.2) if H^a generates a faithful representation of G. The charges of 't Hooft-Polyakov monopoles satisfy also a more stringent condition namely (14.2) where H^a generates a faithful representation of the universal crossing group $\bar{G}$ of G. For example in the SO_3 theory the charge of the original 't Hooft-Polyakov monopole satisfies (14.2) with $H = \tau_3/2$ in the 2×2 representation faithful in SU_2. The distinction reflects the fact that the solution of (14.2) forms a representation of $\pi_1(H)$ for a faithful representation of G, and of $\pi_1(\bar{H}) \subseteqq \pi_1(H)$ if one imposes the restriction to $\bar{G}$. Indeed, $\pi_1(\bar{H})$ contains only those homotopy classes of H, corresponding to paths which are contractable in G.

15. THE CHARGE LATTICE

There exists an elegant representation of the magnetic charges that satisfy the quantization condition (14.2) introduced by Englert and Windey[16).]

Consider a faithful representation of $\bar{G}$ with basis ψ_κ. To each basis vector corresponds an ℓ-dimensional weight vector

$$\vec{H}|\psi_\kappa\rangle = \vec{m}_\kappa|\psi_\kappa\rangle \,. \tag{15.1}$$

The weight vectors have the property that for any root α_i the following condition holds

$$\vec{m}_\kappa\cdot\vec{\alpha}_i = \tfrac{1}{2}n\,|\vec{\alpha}_i|^2 \quad , \qquad n\in Z \,. \tag{15.2}$$

Hence for any element

$$\vec{\tilde{\alpha}} = \sum_{i=1}^{\ell} K_i\,\vec{\alpha}_i/|\vec{\alpha}_i|^2 \qquad K_i \in Z \tag{15.3}$$

of the "inverse root lattice" it follows that

$$\vec{\tilde{\alpha}}.\vec{H}|\psi_\kappa\rangle = \vec{\tilde{\alpha}}\,\vec{m}_\kappa|\psi_\kappa\rangle = \tfrac{1}{2}\,n'|\psi_\kappa\rangle \,,\ n'\in Z \,. \tag{15.4}$$

In other words the solution of the quantisation condition can be expressed as

$$\frac{eg}{4\pi} = \frac{eg^a}{4\pi}H^a = \vec{\tilde{\alpha}}.\vec{H} \,. \tag{15.5}$$

The magnetic charges of nonsingular monopoles correspond to the points of the inverse root lattice of G. For a group with all roots of equal length the charge lattice is isomorphic with the root lattice. If this is not the case one obtains the inverse root system from the root system by interchanging short and long roots. A brief inspection of the Dynkin diagrams tells us that all simple root systems are isomorphic to their inverse except for SO(2N+1) and SP(2N) which get interchanged under inversion.

The use of the charge lattice is not restricted to the case where the residual symmetry group H is the maximal torus. The topologically conserved components of the magnetic charge are associated with separate U_1 factors of H (we ignore the rather exotic possibility of not simply connected nonabelian factors in H). In other words, those generators $\vec{\tilde{\alpha}}_i.\vec{H}$ which do not generate separate U_1 factors of H, correspond to directions in the charge lattice in which the magnetic charge is not conserved.

Brandt and Neri[17] have shown that an initial configuration with some charge $\{K_i\}$,

$$\frac{eg}{4\pi} = \sum_{i=1}^{e} K_i \; \vec{\alpha}.\vec{H}$$

is unstable. It will decay into a configuration with charhe $\{K_i'\}$ with $K_m' = o$ for all m which correspond to generators $\vec{\alpha}_m.\vec{H}$ that do not generate a separate U_1 factor of H. In other words, a monopole configuration minimizes its total charge by reducing the non conserved components. Further stability considerations[18] suggest that the configuration just obtained will in fact decay in a system of separated monopoles with fundamental charges $\frac{eg_i}{4\pi} = \vec{\alpha}_i.H$ $(i = 1,2,\ldots,\ell)$.

16. SOME EXACT SOLUTIONS FOR ARBITRARY G

The construction of the general multimonopole solution of the Bogomolny equations for a theory with some arbitrary gauge group G poses an as yet unsolved problem. In order to simplify the problem one may however impose additional symmetry conditions on the solution. This reduces the number of independent field variables as well as the number of coordinates. Because the Bogomolny equations are expected to be completely integrable one may hope to be able to use some of the solution generating techniques which are available for one and two dimensional integrable nonlinear systems. In fact it may be possible to construct new lower dimensional systems which are completely integrable by imposing extra symmetries on the Bogomolny equations.

A first instance of where exact solutions were obtained is the case of spherical symmetry[19]. One imposes symmetry under a mixed angular momentum operator

$$\vec{J} = \vec{L} + \vec{T} \tag{16.1}$$

where $\vec{L} = -i(\vec{r} \times \vec{\nabla})$ is the usual angular momentum generator and $\vec{T}$ generates some SU_2 subgroup of G. In other words we demand that the effect of a rotation in real space can be undone by a rigid gauge transformation. The solution should then have the following transformation properties:

$$[J_i, A_j] = i\varepsilon_{ijk} A_k \,, \tag{16.2}$$

and

$$[\vec{J}, \Phi] = 0 \,. \tag{16.3}$$

It can be shown that for a particular choice of $\vec{T}$ (corresponding to the principal (maximal) SU_2 embedding) the system reduces to

$$\partial_+ \partial_- \rho_i = \sum_{j=1}^{\ell} K_{ij} \, e^{\rho_j} \qquad (i,j = 1,\ldots,\ell) \,. \tag{16.4}$$

We have allowed for a nontrivial time dependence and introduced the variables

$$z_\pm = r \pm it \,. \tag{16.5}$$

The system (16.4) is apparently completely characterized by the Cartan matrix of the Lie algebra G, defined as

$$K_{ij} = 2\vec{\alpha}_i \cdot \tilde{\vec{\alpha}}_j \,. \tag{16.6}$$

For $G = SU_2$, equation (16.4) is the well known Liouville equation[20]. Imposing $\rho = \rho(r)$ yields static monopole solutions whereas $\rho = \rho(r,t)$ produces instanton solutions. Finally the constraint $\rho = \rho(t)$ gives some generalisation of the Toda chain. (In fact for $G = SU_N$ it is exactly the Toda chain). A general solution in terms of ℓ (= rank G) arbitrary anlytic functions has been constructed[21].

We conclude with a brief discussion of the axially symmetric case[22]. The Bogomolny equations reduce to a nonlinear sigma model in the curved (ρ,z) space. That is,

$$\vec{\nabla}\left[\rho(\vec{\nabla}\mu)\mu^{-1}\right] = 0 \tag{16.7}$$

with $\vec{\nabla} = (\partial_\rho, \partial_z)$ and

$$\mu = g^\dagger g \qquad , \qquad g = g(\rho,z) \in G^* \,. \tag{16.8}$$

The group G^* is some particular noncompact form of G. Because of its definition (16.8) μ is in fact an element of the symmetric (coset) space G^*/K where K is the maximal compact subgroup of G^*.

For the case $G = SU_2$ the system (16.7) is equivalent to the Ernst equation[23] of general relativity. Usually presented in terms of some complex function ε,

$$(\mathrm{Re}\ \varepsilon)\ \vec{\nabla}(\rho\vec{\nabla}\varepsilon) - \rho(\nabla\varepsilon)^2 = 0\ . \tag{16.9}$$

This equation has been studied extensively in the context of classical relativity and an algebraic proliferation of solutions is possible using the Bäcklund transformations of Harrison, Neugebauer and Maison[24]. These have now been used to successfully construct multimonopole solutions[25].

Also in the case where G is arbitrary, a number of Bäcklund transformations of (16.7) have been constructed allowing for an algebraic construction of multimonopole solutions[26].

As for the general solution there are various possible ways to go. Firstly, it is to be expected that the approach due to Atiyah, Ward and others by which the general solution in SU_2 was obtained can be applied. Secondly, it may be possible to extract monopole solutions for the arbitrary group from the general instanton solution as constructed by Atiyah, Dringfeld, Hitchin and Manin[27]. Finally one may try to generalize the Bäcklund transformations for the reduced axially symmetric system to the full Bogomolny equations.

17. MONOPOLES CROSSING A PHASE BOUNDARY

In Grand Unified Theories of the strong, weak and electromagnetic interactions one usually encounters the situation where there are two subsequent breakings

$$G \underset{m_X}{\supset} H_1 \underset{m_A}{\supset} H_2\ , \tag{17.1}$$

characterized by mass scales which differ by many orders of magnitude ($m_X \gg m_A$). Such phases may for example have been realized subsequently in the very early universe. This raises an interesting question. What is the fate of topological excitations in the H_1-phase after the transition to the H_2-phase takes place. At this point we are not just interested in classifying the topological excitations in a given phase, but rather in the relation between topological charges in different phases. Ignoring the dynamical complications of the phase transition we can rephrase the problem as that of a monopole crossing a phase boundary between two domains with H_1 and H_2 symmetry respectively. As we will see the homotopy sequence (8.1) contains the answer[28).

Let us first illustrate the problem with an example which involves the models we have discussed in detail:

$$SO_3 \underset{m_X}{\rightarrow} U_1 \underset{m_A}{\rightarrow} Z . \tag{17.2}$$

In the $H_1 = U_1$ phase we have the familiar 't Hooft-Polyakov monopole. This monopole will be confined in the $H_2 = Z$ phase, where we have the Nielsen-Olesen magnetic fluxtube. That is, the H_1 radial magnetic field which has a well defined meaning for $r > 1/m_X$ bends over in a fluxtube with a diameter $d \simeq 1/m_A \gg 1/m_X$. The topological magnetic charge of the U_1 phase is changed to the topological magnetic flux of the Z phase. What is the situation in general?

The answer is provided by the following identity taken from the homotopy sequence,

$$\mathrm{Im}\left[\pi_1(H_2) \rightarrow \pi_1(H_1)\right] = \mathrm{Ker}\left[\pi_1(H_1) \rightarrow \pi_1(H_1/H_2)\right] . \tag{17.3}$$

We recall the physical meaning of the homotopy groups:

- $\pi_1(H_1)$, <u>all</u> top. conserved magnetic charges in H_1-phase
- $\pi_1(H_2)$, <u>all</u> top. conserved magnetic charges in H_2-phase
- $\pi_1(H_1/H_2)$, <u>new</u> top. conserved magnetic flux in H_2-phase.

The relation (17.3) makes a statement about the elements of $\pi_1(H_1)$. The homomorphism $\pi_1(H_2) \rightarrow \pi_1(H_1)$ is associated with lifting closed paths in H_2 to H_1 ($\supset H_2$). The H_1 monopole charges associated with the image of this homomorphism correspond to the trivial element of $\pi_1(H_1/H_2)$. Consequently their magnetic flux is not squeezed into a tube upon entering the H_2-phase. One of three things may happen to these monopoles, they

1) survive unchanged,
2) are converted to H_2 monopoles,
3) decay into the vacuum.

The elements of $\pi_1(H_1)$ which are not covered by the homomorphism $\pi_1(H_2) \rightarrow \pi_1(H_1)$ correspond to H_1 monopoles that will be confined in the H_2-phase.

18. INSTANTONS IN FOUR DIMENSIONS

Instantons[1)] are solutions to the pure Yang-Mills equations

$$D_\mu F^a_{\mu\nu} = 0 \tag{18.1}$$

with finite Euclidean action. The requirement of finite action imposes the condition that the gauge potentials go to a pure gauge at $|x| \to \infty$

$$A_\mu(|x| \to \infty) = i(\partial_\mu g)\, g^{-1} , \quad g \in G . \tag{18.2}$$

This allows for topologically nontrivial boundary conditions in R^4 because

$$g : S^3_x \to G \tag{18.3}$$

and

$$\pi_3(G) = Z \tag{18.4}$$

for any compact simple Lie group. A map from S^3_x into G can always be continuously deformed into a map of S^3_x into an SU_2 subgroup of G. Then (18.4) follows because the group manifold of SU_2 is S^3. The integer topological charge (or Pontryagin index) is given by

$$\nu = \frac{e^2}{32\pi^2} \int F\tilde{F}\, d^4x , \tag{18.5}$$

where the dual $\tilde{F}$ of F is defined as

$$\tilde{F}_{\mu\nu} = \tfrac{1}{2} \varepsilon_{\mu\nu\alpha\beta} F_{\alpha\beta} . \tag{18.6}$$

The total action of a configuration with some value of ν can be bounded below using Schwartz inequality

$$\left[\int d^4x\, FF \int d^4x\, \tilde{F}\tilde{F}\right]^{\frac{1}{2}} \geqslant \left| \int d^4x\, F\tilde{F} \right| . \tag{18.8}$$

The equal sign applies if the field strength is either selfdual or anti selfdual

$$F_{\mu\nu} = \pm \tilde{F}_{\mu\nu} . \tag{18.9}$$

Solutions to (18.9) are solutions to (18.1) due to the Bianchi identity $D_\mu \tilde{F}_{\mu\nu} \equiv o$.

The (anti) selfduality condition has drawn the attention of physicists and mathematicians alike. Belavin, Polyakov, Schwartz and Tyapkin suggested the problem[29)] in 1975 and produced the $\nu = 1$ solution. The solution was generalized for the case $G = SU_2$ subsequently by Witten, 't Hooft and Jackiv, Rebbi and Wohl[30)]. Using methods from algebraic geometry Atiyah, Hitchin, Drinfeld and Manin[27)] succeeded in obtaining the general solution of (18.9) for an arbitrary group by reducing it to a problem of linear algebra (1978). It is beyond the scope of this lecture to enter the formalism which produced this beautiful result. We mention that it has been shown[31)] that for any compact Euclidean spacetime all finite action configurations are (anti) selfual (at least if $G = SU_2$). We finally note that the problem of static solutions to the selfduality equations is equivalent to obtaining the multimonopole solutions of the Bogomolny equations. Indeed, the selfuality equations constitute one of the most exquisite challenges in contemporary mathematical physics: A rich class of nonlinear partial differential equations in four dimensions of great physical interest, which seem to be completely integrable.

19. VACUUM TUNNELING

Let us consider as the boundary of R^4 a cylinder in the $x_4 = it$ direction with boundaries $S^3_\pm$ for $x_4 = \pm \infty$). Any configuration with finite pontryagin index ν can be brought into a gauge where $g = 1$ (i.e. $A_\mu = o$) for $|\vec{x}| \to \infty$. In such a gauge we may compactify the space R^3 to S^3 for any value of x^4. Moreover, since for $x_4 = \pm \infty$ the gauge potentials go to a pure gauge we may calculate the winding numbers $n_\pm$ by integrating over $S^3_\pm$. These obey the relation

$$\nu = n_+ - n_- .$$

Of course only ν has a gauge invariant meaning. The homotopy argument implies that the n_- vacuum cannot continuously be deformed into the n^+ vacuum. The continuous field configuration which interpolates between n_- and n_+ can therefore not be a pure gauge everywhere. Its action should be nonzero. Indeed, the instanton is exactly the configuration of minimal action which connects the two vacua with different winding numbers[32)].

In physics one is interested in computing vacuum tunneling amplitudes. In field theory this is most easily done by calculating the matrix element of the time evolution operator as a Euclidean path integral.

$$\lim_{t\to\infty} \langle n+\nu | e^{-i\mathcal{H}t} | n \rangle =$$

$$= \int \left[DA_\mu\right]_\nu e^{-S(A_\mu)_\nu} \simeq C\, e^{-\frac{8\pi^2}{e^2}|\nu|} \quad . \tag{19.1}$$

Here $\mathcal{H}$ is the Hamiltonian, S the classical Euclidean action. The integral is over all configurations with Pontyagin index ν [modulo gauge equivalence] and can be performed approximately by perturbation theory around the instanton configuration(s) with minimal action. The exponent is just the action of the instanton and C is a properly defined fluctuation determinant. It is important to note that the amplitude is of order $\exp(-1/e^2)$ assuming $C \neq o$ and therefore can never be recovered by summing the perturbation series in the coupling constant e.

In a situation where there is tunneling between degenerate groundstates $|n\rangle$ one needs to construct some new groundstate ψ,

$$|\psi\rangle = \sum_n c_n |n\rangle \tag{19.2}$$

which is stable under tunneling. In order to compute the coefficients c_n we note that there exists a gauge transformation which changes the winding number by one unit,

$$\Omega(x)|n\rangle = |n+1\rangle \quad . \tag{19.3}$$

The transformation $\Omega(x)$ is of course not homotopic to the identity. Nevertheless Ω is a symmetry of the Hamiltonian, i.e. $[\Omega,\mathcal{H}] = o$. Furthermore we have that $\Omega^\dagger\Omega = 1$. So the lowest order beyond the classical one we can diagonalize $\mathcal{H}$ and Ω simultaneously by

$$\Omega|\psi\rangle = e^{i\theta}|\psi\rangle \tag{19.4}$$

so that

$$c_n = e^{i\theta n} \quad . \tag{19.5}$$

The angle θ is a new, arbitrary parameter of the theory, and we should calculate amplitudes using the coherent groundstates or θ-vacua

$$|\theta\rangle = \sum_n e^{i\theta n} |n\rangle \ . \tag{19.6}$$

One has for example,

$$\lim_{t\to\infty} \langle \theta' | e^{-iHt} | \theta \rangle = \delta(\theta-\theta') \, I(\theta) \ , \tag{19.7}$$

with

$$I(\theta) = \int DA_\mu \ \exp(-S(A_\mu) - S_\theta) \ , \tag{19.8}$$

where

$$S_\theta = i\theta\nu = \frac{i\theta e^2}{32\pi^2} \int F\tilde{F} \, d^4x \ . \tag{19.9}$$

Note that the theory with $\theta \neq o$ violates parity and time reversal symmetry. This fact poses a new problem in QCD the theory of the strong interactions which are invariant under parity. The question is now: Why does θ_{QCD} vanish? At the other hand we should mention that the existence of the θ parameter has solved the long standing problem of a "missing" Goldstone particle in the spectrum of strongly interacting matter[33].

20. EPILOGUE

We have given an exposé of a variety of physical manifestations of the nontrivial topology inherent to gauge theories. There are some important issues however, which were not covered in these lectures. For example the problem of electric-magnetic duality in nonabelian gauge theories. In particular 't Hooft's exact duality relations which were derived by studying the Yang-Mills equations on the hypertorus T^4. We hope to have succeeded in wetting the reader's appetite, and encourage the interested reader to consult other reviews as well as the original literature in order to find answers to questions this presentation may have raised but failed to clarify.

REFERENCES

1. There are many reviews on this subject, we mention a few:
   On topological solitons in general:
   S. Coleman, Lectures at the 1975 International School of Subnuclear Physics, Ettore Majorana, Erice, Sicily.
   On Magnetic Monopoles:
   P. Goddard and D. Olive, Rep. Progr. Phys. 41 (1978) 1357.
   On Instantons:
   S. Coleman, Lectures at the 1977 International School of Subnuclear Physics, Ettore Majorana, Erice, Sicily.
2. D. Saint-James, G. Sarma and E.J. Thomas, *Type II Superconductivity*, Pergamon Press (1969).
3. H.B. Nielsen and P. Olesen, Nucl. Phys. B61 (1973) 45.
4. E. Weinberg, Phys. Rev. D19 (1979) 3008.
5. A. Jaffe and C. Taubes, *Vortices and Monopoles*, Birkhauser, Boston (1980).
6. G. 't Hooft, Nucl. Phys. B153 (1979) 141; Commun. Math. Phys. 81 (1981) 267.
7. D.J. Gross, R.D. Pisarski and L.G. Yaffe, Rev. Mod. Phys. 53 (1981) 43.
8. F.A. Bais, Nucl. Phys. B170 [FS1] (1980) 32;
   V. Poenaru and G. Toulouse, J. de Phys. (Paris) 38 (1977) 887.
9. G. 't Hooft, Nucl. Phys. B79 (1974) 276;
   A.M. Polyakov, JETP Lett. 20 (1974) 194.
10. B. Julia and A. Zee, Phys. Rev. D11 (1975) 2227.
11. E.B. Bogomolny, Sov. J. Nucl. Phys. 24 (1976) 449.
12. M.K. Prasad and C.M. Sommerfield, Phys. Rev. Lett. 35 (1975) 760.
13. E. Weinberg, Nucl. Phys. B167 (1980) 500.
14. A. Jaffe and C. Taubes, *Vortices and Monopoles*, Birkhauser, Boston (1980).
15. R.S. Ward, Phys. Lett. 61A (1977) 81;
    M.F. Atiyah and R.A. Ward, Comm. Math. Phys. 55 (1977) 117.
16. F. Englert and P. Windey, Phys. Rev. D14 (1976) 2728;
    P. Goddard, J. Nuyts and D. Olive, Nucl. Phys. B125 (1977) 1.
17. R. Brandt and F. Neri, Nucl. Phys. B161 (1979).
18. F.A. Bais, Phys. Rev. D18 (1978) 1206;
    E. Weinberg, Nucl. Phys. B167 (1980) 500.
19. F.A. Bais and H.A. Weldon, Phys. Rev. Lett. 41 (1978) 601;
    D. Wilkinson and F.A. Bais, Phys. Rev. D19 (1979) 2410;
    F.A. Bais, Proceedings of the International Workshop on High Energy Physics and Field Theory, Serpukhov (1979).

20. E. Witten, Phys. Rev. Lett. 38 (1977) 121.
21. A.N. Leznov and Saveliev, Lett. in Math. Phys. 3 (1979) 489; Comm. Math. Phys. 74 (1980) 111.
22. F.A. Bais and R. Sasaki, Nucl. Phys. (to be published).
23. F.J. Ernst, Phys. Rev. 167 (1968) 1175.
24. D. Maison, Phys. Rev. Lett. 41 (1978) 521;
B.K. Harrison, Phys. Rev. Lett. 41 (1978) 1197;
G. Neugebauer, J. Phys. A12 (1979) L67.
25. P. Forgács, Z. Horváth and L. Palla, Phys. Rev. Lett. 45 (1980) 505;
Phys. Lett. 99B (1981) 232.
26. F.A. Bais and R. Sasaki, Nucl. Phys. B (to be published).
27. M.F. Atiyah, N.J. Hitchin, V.G. Drinfeld and Yu. I. Manin, Phys. Lett. 65A (1978) 185.
28. F.A. Bais, Phys. Lett. 98B (1981) 437.
29. A.A. Belavin, A.M. Polyakov, A.S. Schwarz and Y.S. Tyupkin, Phys. Lett. 59B (1975) 85.
30. E. Witten, Phys. Rev. Lett. 38 (1977) 121;
G. 't Hooft, (unpublished);
R. Jackiw, C. Nohl and C. Rebbi, Phys. Rev. D15 (1977) 1642.
31. J.P. Bourguignon, H.B. Lawson and J. Simons, Proc. Nat. Acad. Sci 76 (1979) 1550.
32. R. Jackiw and C. Rebbi, Phys. Rev. Lett. 37 (1976) 172;
C. Callen, R. Dashen and D. Gross, Phys. Lett. 63B (1976) 334.
33. G. 't Hooft, Phys. Rev. Lett. 37 (1976) 8;
Phys. Rev. D14 (1976) 3432.

PARTICLES, FIELDS AND QUANTUM THEORY

P.J.M. BONGAARTS

Particles, fields and quantum theory.

This conference brings together mathematicians and theoretical physicists, people interested and knowledgeable in subjects that have unfortunately become more and more separeted. It may therefore be a good idea to add to the regular, rather advanced talks a few talks of a more general and introductory character. Such talks will be quite elementary to one half of the audience, but may at the same time provide useful and even necessary background information to the other half. It is in this spirit that I shall discuss particles, fields and quantum theory. For the sake of coherence, an admittedly very simplified historical scheme will be followed.

1. Classical physics.

According to the views dominating physics towards the end of the nineteenth century, the two basic constituents of physical reality where matter and the electromagnetic field, described by two fundamental theories, Newtonean mechanics and Maxwell's theory of electromagnetism. This picture, sometimes called classical physics, can be summarized as follows:

There is matter, consisting of particles. Motion is described by classical mechanics: A system of N point-particles has N time-dependent position vectors $\vec{r}_1(t),\ldots,\vec{r}_N(t)$, satisfying a system of second order ordinary differential equations, essentially Newton's equations

$$m_j \frac{d^2}{dt^2}\vec{r}_j(t) = \vec{F}_j(\vec{r}_1(t),\ldots,\vec{r}_N(t)) \qquad (1)$$

$$(j=1,\ldots,N; \quad m_j \text{ mass of } j^{th} \text{ particle})$$

The forces $\vec{F}_j$ are usually derivable from a potential function. It is then profitable to study the equations as the Euler-Lagrange equations of a variational problem. They can also be transformed into first order equations (Hamilton's equations).

There is also the electromagnetic field, associated with radiation. It consists of quantities, defined at every point of space and dependent on time, the vector fields of electric and magnetic field strength $\vec{E}(\vec{r},t)$ and $\vec{B}(\vec{r},t)$. Their properties in empty space, in particular their time evolution, are determined by Maxwell's equations, a set of linear first order partial differential equations. One often uses auxiliary field quantities, scalar and vector potentials $\phi(\vec{r},t)$ and $\vec{A}(\vec{r},t)$, from which the physical quantities $\vec{E}$ and $\vec{B}$ can be derived and which are mathematically more convenient.

Electromagnetic fields act as forces on charged particles, moving particles in turn generate fields. To the separate equations of classical mechanics and Maxwell theory are therefore added somewhat complicated connecting equations. The result is an integrated description of matter and electromagnetic fields, which forms, apart from gravitational forces, the basis of classical physics.

2. Transitional period.

Around 1900 the majority of known physical phenomena were extremely well described by classical physics. There remained however a few unexplained and paradoxical effects, at first considered as of marginal importance, but gradually, with increasing experimental knowledge of the sub-microscopic world, recognized as challenges to the essential features of classical physics. Well-known is of course the problem of discrete atomic spectra. An atom, such as the hydrogen atom, emits radiation only in sharply defined, discrete frequencies. There is no way to understand this from classical mechanics and Maxwell theory. There was a long period of confusion and struggle, in which various imaginative ad hoc proposals, such as Bohr's model for the hydrogen atom were tested. Finally, in 1925, new ideas appeared that made it possible to put things together in a single framework and in a very short time a new and complete mechanical theory emerged, called quantum mechanics.

3. Quantum mechanics.

Quantum mechanics was a new fundamental description of mechanical systems, taking the place of classical mechanics and relegating this to the rôle of an approximative theory, still sufficiently accurate for many macroscopic situations. The names of Born, Heisenberg, Schrödinger and others are associated with its beginning in 1925.

The description of a system of point particles in terms of quantum mechanics is very different from that by classical mechanics.

The first important physical aspect to be mentioned is the fact that classical quantities like position and moment still make sense and can be measured, but no longer simultaneously. Consequently such a concept as the orbit of a moving particle becomes meaningless. Secondly the theory only predicts the results of experiments in terms of probabilities. This lack of determinacy is not due to imperfections in experimental methods, but is considered to be a fundamental property of matter, that becomes particularly noticable on a sub-microscopic scale. The mathematics of quantum mechanics is also very different. The physical principles of quantum mechanics are expressed in a mathematical formalism that lacks the immediate intuitive appeal of classical mechanics. Nevertheless it is transparent and essentially simple. This can be best appreciated by leaving this historical discussion for a moment and looking at the general formalism of modern quantum theory of which quantum mechanics is now a special example.

In quantum theory a physical system is described by a complex Hilbert space $\mathcal{H}$ and self-adjoint operators A in $\mathcal{H}$. Vectors ψ in $\mathcal{H}$, of unit length, describe possible states of the system. (With the understanding that ψ and $e^{i\alpha}\psi$ mean the same state.) Self-adjoint operators A represent physical observables, the quantities that can be measured by experiment, such as position, momentum, energy, etc. If the state vector ψ is known at a certain instant of time, one may derive from it probabilities for the possible outcomes of the measurement of an arbitrary observable A, in case such a measurement should take place, by the following relation

$$(\psi, e^{iuA}\psi) = \int_{-\infty}^{+\infty} e^{iu\alpha} dF(\alpha) \tag{2}$$

The left hand side is, as function of the real variable u, a function of positive type in the sense of Bochner's theorem, due to the self-adjointness of A. (It is also normalized to unity in u=0 because $||\psi|| = 1$.) It is therefore the Fourier-transform of a unique probability measure, of which $F(\alpha)$ is the distribution function. Slightly less rigorously, (2) means that $F(\alpha)$ is the distribution with expectation and higher moments given by

$$\overline{\alpha^n} := \int_{-\infty}^{+\infty} \alpha^n dF(\alpha) = (\psi, A^n\psi) \qquad (n=1,2,3,\ldots) \tag{3}$$

By using the spectral theorem for A one shows that this probability distribution is concentrated on the spectrum of the operator A. This has an important physical interpretation: The only values that a physical quantity can possibly have, are numbers from the spectrum of the corresponding operator. Some quantities that in classical mechanics may assume a continuous range of values, such as the energy of a harmonic oscillator or the hydrogen atom, have to be represented in quantum mechanics by operators with discrete spectrum. This explains the discrete frequencies in atomic radiation.

For commuting self-adjoint operators $A_1,\ldots,A_n$, an obvious generalization of (2) defines a joint probability distribution $F(\alpha_1,\ldots,\alpha_n)$, for each state vector ψ. For non-commuting operators such joint distributions do not exist. This appears in the physical interpretation as the fact that quantities corresponding with non-commuting operators, like e.g. position and momentum of a particle, cannot be measured simultaneously.

The dynamical behaviour of the system is specified by a particular self-adjoint operator $\hat{H}$, the Hamiltonian, usually, but not always, the operator that represents the energy as an observable. It generates a one-parameter group of unitary operators $U_t = e^{-\frac{it}{\hbar}H}$. ($\hbar$ is Planck's constant, divided by 2π). There are two different, but physically equivalent ways of employing the operators U_t:

The U_t can be used to make the state vectors time-dependent, according to $\psi_{t_o+t} = U_t\psi_{t_o}$. This is called the Schrödinger picture of quantum theory. The basic dynamical equation is in this case

$$-\frac{\hbar}{i}\frac{d}{dt}\psi_t = \hat{H}\psi_t \tag{4}$$

Sometimes it is more convenient to realize the time evolution by considering the observables as time-dependent: $A_{t_o+t} = U_t^{-1}A_{t_o}U_t$. This is called the Heisenberg picture. Its dynamical equation is the operator equation

$$\frac{d}{dt}A_t = \frac{i}{\hbar}[H,A_t] \qquad ([H,A_t] \text{ stands, of course, for } HA_t - A_tH) \tag{5}$$

Because of the identity $(U_t\psi, AU_t\psi) = (\psi, U_t^{-1}AU_t\psi)$ both pictures obviously lead to the same probabilistic statements and are therefore physically completely equivalent.

We return now to our historical discussion and to the special case of quantum mechanics.

For the description of the simplest case, a single particle in a potential $V(\vec{r})$, ($\vec{r}=(x^1,x^2,x^3)$), the Hilbert space $\mathcal{H}$ is taken to be $L_2(\mathbb{R}^3, dx^1,dx^2,dx^3)$, the space of square integrable complex valued functions $\psi(\vec{r})$, called wave functions. The basic observables are position and momentum:

One has three position operators:

$$\hat{x}^j : \psi(x^1,x^2,x^3) \longmapsto x^j\psi(x^1,x^2,x^3) \qquad j=1,2,3 \tag{6}$$

From (3), or rather the generalization to a set of commuting operators mentioned in the text, one derives that the joint probability for simultaneous measurement of the co-ordinates of position x^1, x^2, x^3, in a state given by $\psi(x^1, x^2, x^3)$, is simply $dF(x^1, x^2, x^3) = |\psi(x^1, x^2, x^3)|^2 dx^1 dx^2 dx^3$.

There are also three momentum operators

$$\hat{p}_j : \psi(x^1,x^2,x^3) \longmapsto \frac{\hbar}{i}\frac{\partial}{\partial x^j}\psi(x^1,x^2,x^3) \qquad j=1,2,3 \tag{7}$$

The generalization of (3) gives as joint probability distribution for the components of momentum $dF(p_1,p_2,p_3)=|\phi(p_1,p_2,p_3)|^2 dp_1 dp_2 dp_3$, with ϕ the Fourier transform of ψ. Clearly not all operators $\hat{x}^j$, $\hat{p}_k$ commute among each other, one has in fact

$$[\hat{p}_j, \hat{x}^k] = \frac{\hbar}{i}\delta_j^{\,k} \tag{8}$$

This expresses the impossibility of a complete simultaneous measurement of position and momentum.

The operators for other observables, such as energy and angular momentum are expressions in the $\hat{p}_j$ and $\hat{x}^k$, the same expressions as for the classical quantities.

For the time evolution one prefers in quantum mechanics the Schrödinger picture. The time dependent wave function $\psi(x^1,x^2,x^3,t)$ satisfies a partial differential equation, known as the Schrödinger equation, which is the explicit form of the dynamical equation (4), with as Hamiltonian operator the operator for the total energy

$$\hat{H} = -\frac{\hbar}{2m}\sum_{j=1}^{3}\frac{\partial^2}{(\partial x^j)^2} + V(x^1,x^2,x^3) \tag{9}$$

A system of N particles is described in a similar way by a Hilbert space of wave functions $\psi(\vec{r}_1,\ldots,\vec{r}_N)$, with 6N operators for positions and momenta, multiplication and differentiation operators with respect to the separate position co-ordinates.

The most important practical problems in quantum mechanics are to find, for specific potentials V, the eigenvalues, in general the spectrum, of $\hat{H}$ (possible energy values) and to solve the Schrödinger equation for given initial-time wave functions. (Transition probabilities) Except in a few simple cases these problems can only be solved approximately. (Perturbation theory)

A few final remarks on quantum theory in general:

a. The Hilbert spaces used for different physical systems are separable and therefore all isomorphic. The various cases differ in the sets of operators used as observables. This leads to different concrete realizations of Hilbert space, for instance in terms of different function spaces.

b. There is a two-way relation between classical and quantum theories. On one hand the appropriate quantum description for a given physical system is usually obtained by assigning in a certain systematic way a concrete Hilbert space and operators to the main elements of the classical description, cast in some standard form. Such quantization procedures are useful even in those cases where no classical theories are physically meaningful, and where fictitious "classical" theories have been introduced solely for this purpose. This happens in most of sub-nuclear physics, as will be discussed. On the other hand, in situations where the classical theories still give accurate results, such as in the domain of macroscopic mechanics, it is possible to obtain the classical theory, or at least its numerical results by means of a limit procedure within quantum theory. (Classical limit)

c. The accepted standard form of quantum theory is the Hilbert space formalism that we have discussed. There is another formalism that looks rather different, is supposed to the equivalent and has recently acquired considerable popularity. (Path-integral quantization)

4. Transitional period.

The introduction of quantum mechanics led to an at least qualitatively satisfactory understanding of atomic physics. Further improvements were however needed. An atom was seen as consisting of a nucleus with a number of electrons moving around it; everything kept together by electro-magnetic forces. The description of such a system by quantum mechanics combined with classical Maxwell theory could only be a first step. The next thing to do was to replace also the classical theory of electro-magnetism by a quantum version. Such a theory, the first example of a quantum field theory, was set up by Dirac, already in 1927, and developed further by him and others in subsequent years.

In this theory of the electro-magnetic field, the basic quantities are still the electric and magnetic field strengths. In accordance with the general ideas of quantum theory, they appear however as operator fields $\hat{\vec{E}}(\vec{r},t)$ and $\hat{\vec{B}}(\vec{r},t)$, or $\hat{E}_j(\vec{r},t)$ and $\hat{B}_j(\vec{r},t)$, $j=1,2,3$, i.e. systems of self-adjoint operators, indexed by the position variable $\vec{r}=x^1,x^2,x^3$ and time variable t, and acting in a Hilbert space of state vectors $\mathcal{H}$. The dependence on t means that in this case the Heisenberg picture of quantum theory is used, in which the observables, not the state vectors, change in time. The dynamical behaviour is therefore determined by the operator equation (5), which turns out to be noting else but the system of Maxwell's equations, understood as a set of partial differential equations for operator-valued functions of $\vec{r}$ and t.

The electro-magnetic field is a system with an infinite number of degrees of freedom. In quantum mechanics of an N-particle system one has 6N operators for positions and momenta, while here there is an infinite set of basic operators, the field strength operators at every point of space. Consequently the explicit realization of the Hilbert space $\mathcal{H}$ is much more complicated. There does exist a realization of $\mathcal{H}$ as an L_2 space. It is a space of functionals, square integrable with respect to a measure in an infinite dimensional space. Because of the mathematical subtleties involved and for other reasons, its usefulness has so far been rather restricted.

A more convenient, though still fairly complicated realization of $\mathcal{H}$ has the form

$$\mathcal{H} = \sum_{n=0}^{\infty} \oplus (\otimes^n \mathcal{H}^{(1)})_s \qquad (10)$$

and is an infinite direct sum of symmetrized n-fold tensor product spaces, constructed from a single Hilbert space $\mathcal{H}^{(1)}$. It is called Fock space and is connected with a very important physical aspect of the radiation field: Under certain experimental conditions electro-magnetic radiation, in particular light exhibits a discrete, corpuscular behaviour. This was known from 1900 onwards. The particles for light were called photons and their existence was another one of the challenges to the general picture of classical physics. In the formalism for the quantized electro-magnetic fields, operators representing the measurement of the presence of such particles arise naturally. The realization (10) is associated with these operators.

Because of its importance for quantum field theory in general, we shall explain this, although necessarily in a very schematic and simplified manner.

In classical electro-magnetism, one is often, except in purely static situations, not so much interested in the measurement of the field strengths themselves, as in the determination of intensities of radiation, in certain directions and at certain specific frequencies. If we represent the electro-magnetic field by the potential $A(\vec{r},t)$, of which we suppress the vector character for the sake of simplification, this may be written, by means of a complex Fourier expansion, in terms of plane waves as

$$A(\vec{r},t) = \int (a(\vec{k})e^{i(\vec{k}.\vec{r}-\omega t)} + a^*(\vec{k})e^{-i(\vec{k}.\vec{r}-\omega t)})d\vec{k} \qquad (11)$$

with wave vectors $\vec{k}$, frequencies $\nu = \frac{\omega}{2\pi}$, depending on $\vec{k}$ as $\omega(\vec{k})=c|\vec{k}|$, c the velocity of light. The intensity of radiation, in the direction given by $\vec{k}$, at the frequency ν is then proportional to $n(\vec{k})=a^*(k)a(\vec{k})=|a(\vec{k})|^2$. (If the field does not propagate freely, the doefficients $a(\vec{k})$ and therefore $n(\vec{k})$ depend also on t.)

In the quantized theory of electro-magnetism we may also describe the field by a potential, which is then an operator valued function $\hat{A}(\vec{r},t)$. It has a similar expansion

$$\hat{A}(\vec{r},t) = \int (\hat{a}(\vec{k})e^{i(\vec{k}.\vec{r}-\omega t)} + \hat{a}^*(\vec{k})e^{-i(\vec{k}.\vec{r}-\omega t)})d\vec{k} \qquad (12)$$

The Fourier coefficients $\hat{a}(\vec{k})$ form a system of operators, with the $\hat{a}^*(\vec{k})$ as their hermitian adjoints. The operators $\hat{n}(\vec{k})=\hat{a}^*(\vec{k})\hat{a}(\vec{k})$ are again self-adjoint. Their spectrum can be found from the general properties of the field operators $A(\vec{r},t)$. It is discrete and, with an appropriate normalization convention in (12), consists, for every $\hat{n}(\vec{k})$, of the numbers 0,1,2,3,.... If the $\hat{n}(\vec{k})$ still describe the intensity of radiation in some sense, this indicates that for each frequency, radiation only occurs in discrete fixed amounts. If one further knows that, in quantum theory, there are good reasons to interprete $\hbar\vec{k}$ as a momentum $\vec{p}$, then it is no longer far-fetched to identify these "quanta" of radiation as photons, <u>particles</u> with momenta $\vec{p}=\hbar\vec{k}$, associated with the quantized electro-magnetic field. In the standard sense of quantum theory, as explained in 3., each operator $\hat{n}(\vec{k})$ represents a physical observable. This observable is the number of photons with momentum $\vec{p}=\hbar\vec{k}$, present in a given state ψ of the field system.

This particle interpretation has shown it self to be very satisfactory and important, not only for the electro-magnetic field, but quite generally in quantum field theory, as will be discussed. It should be stressed that this concept of a <u>particle</u> connected with a <u>quantum field</u> is not only different from the concept of a particle in classical mechanics, but also quite different from that in <u>quantum mechanics</u>. For one thing the number of particles is a true quantum variable. If the field is not free, the operators $\hat{n}(\vec{k})$ will, as operators in the Heisenberg picture, depend on time; i.e. the number of particles may change in time. This means that there is <u>creation</u> and <u>annihilation</u> of particles.

The operators $\hat{a}(\vec{k})$ and $a^*(\vec{k})$, called annihilation operators and creation operators respectively, play an important rôle in the formalism of the quantized electro-magnetic field, and in quantum field theory in general. They generate the Fock space realization (10) of $\mathcal{H}$. In $\mathcal{H}$ the subspaces $(\otimes^n \mathcal{H}^{(1)})_s$, consist of the state vectors for which, with probability one, the total number of particles, described by an operator $\hat{N}$, obtained by integrating the operators $\hat{n}(\vec{k})$ over all $\vec{k} \in \mathbb{R}^3$, is exactly n. For n=0, one has the space $\otimes^0 \mathcal{H}^{(1)}$, which is one dimensional; it corresponds to the <u>vacuum state</u>, the state in which no particles of any kind are present. The operators $\hat{a}^*(\vec{k})$ go from $(\otimes^n \mathcal{H}^{(1)})_s$ to $(\otimes^{n+1} \mathcal{H}^{(1)})_s$, the $\hat{a}(\vec{k})$ in the reverse direction. They can be used to generate the whole space $\mathcal{H}$ from the vacuum state.

The combination of quantum mechanics and the quantum theory of the electromagnetic field made a fully quantized description of atomic systems possible. This was a distinct improvement, nevertheless the result remained unsatisfactory. The main reason for this is the theory of relativity, which we have neglected so far in our historical discussion. Experimental progress was made by employing higher energies for atomic collisions, and it is precisely at high energies, associated with high velocities, that relativistic effects become important. Maxwell's equations, both in their classical form and in the quantized operator version, are such that this can be taken into account. The Schrödinger equation of quantum mechanics is however definitely non-relativistic.

For this reason Dirac proposed in 1928 a new quantum mechanical equation that had better properties with respect to Lorentz transformations, the main criterion for compatibility with the theory of relativity. Its simplest form, without interaction terms, is

$$(i\gamma^{\mu}\partial_{\mu} - m)\psi(x) = 0 \tag{13}$$

In this equation $\psi(x)$ is a 4-component complex function of space and time, the γ^{μ}, $\mu=0,1,2,3$, are certain 4x4 matrices, m the mass of the particle considered. The equation is written in modern relativistic notation, i.e. with $x=\{x^{\mu}\}$, $\mu=0,1,2,3$. $x^{0}=t$, $\partial_{\mu}=\frac{\partial}{\partial x^{\mu}}$, summation convention for μ, suppressing of matrix indices and in units such that $\hbar=c=1$.

As a relativistic version of the Schrödinger equation, and with the obvious interpretation of $\psi(x)$ as a quantum mechanical wave function for a one-particle system, the Dirac equation had a certain succes, but then ran into serious and quite fundamental problems, the nature of which cannot be discussed here.

It took a long period to realize that the only satisfactory way of dealing with these difficulties was to give the equation a radically different meaning. Looking back, the change in point of view involved seems quite clear, although in fact it took place gradually, first in an implicit way and surrounded by conceptual confusion. To understand what happened it may be helpful to compare the situation with that of electromagnetism. There, a particle concept, appropriate for an experimentally known particle-like behaviour of the field, emerged quite naturally from the quantization of the field, as we have indicated. Here in some sense a converse idea is taken up. It is that particles like electrons, protons, etc., should also be understood as particles associated with an underlying quantum field, even if no corresponding classical field has been known before. This results in the notion that $\psi(x)$ in (13) is not a wave function, but an operator field, a system of operators acting in a Hilbert space of state vectors, a Fock space of the general form (10), describing states with various numbers of particles.

With this new description of a system of particles, quantum mechanics has in turn, like classical mechanics earlier, lost its position as a fundamental theory and has become an often useful approximative description.

The Dirac equation (13) is an operator evolution equation in the sense of (5), as such it can be coupled to the operator equations of the quantized electromagnetic field. This gives and integrated quantum field theory, describing systems of variable numbers of, for example, electrons and photons, in interaction with each other. This theory has been overwhelmingly succesful. It is called quantum electrodynamics.

5. Quantum electrodynamics.

In its final form, which it reached in the early fifties, quantum electrodynamics is given by an, at least at first sight, simple set of non-linearly coupled field equations for the operator fields $\psi(x)$ and $A_\mu(x)$

$$\begin{aligned} (i\gamma^\mu\partial_\mu - m)\psi &= e\gamma^\mu A_\mu \\ \partial^\mu\partial_\mu A_\nu &= e\tilde{\psi}\gamma_\nu\psi \end{aligned} \qquad (19)$$

In this, ψ denotes the Dirac quantum field, describing particles like electrons and positrons, with mass m and electric charge e, $\tilde{\psi}$ denotes a particular type of hermitian adjoint of ψ and A_μ is the vector potential for the quantized electromagnetic field, written in a relativistic way as a 4-vector. (From here on we omit the ^, which we have been using to distinguish operators from the corresponding classical quantities, and which is not a standard notation.)

In applying quantum electrodynamics, no attempts are ever made to solve the equations (19) as such. Instead of this, one looks at (19) as the set of Euler-Lagrange equations associated with a 4-dimensional variational problem, in which an integral, the action $\int \mathcal{L}(\psi,\partial_\mu\psi,A_\mu,\partial_\nu A_\mu)d^4x$ is required to be stationary under variations of the fields ψ and A_μ. The expression for the action is the real heart of the theory. It is the basis for a formal power series expansion, from which transition probabilities for the occurrence of all the different processes of creation, annihilation and just simple collision for the various particles involved, can be calculated, in successive orders of approximation. The effectuation of this was for a long time held up by the disturbing fact, that these calculations lead to complicated many-dimensional integrals, that are almost all divergent. By the efforts of Feynman, Dyson, Schwinger and other, some time before 1950, a general method was developed to remove these divergences and to obtain finite numerical results. This method, although systematic and effective, is heuristic and not really well understood. It is called <u>renormalization</u>. After this, quantum electrodynamics has had an unmitigated succes as a physical theory. It has stood up to experimental tests of the greatest precision, down to the present day.

This stands in strong contradiction to its mathematical status. Quantum mechanics, although not always presented that way, is a rigorous mathematical theory, based on standard Hilbert space methods, with the spectral theory of self-adjoint operators as central part. For quantum electrodynamics, and for quantum field theory in general, such a mathematical basis, except in the relatively uninteresting case of free fields, does not exist in any rigorous sense. All operator expressions, again except those for free fiels, are highly formal. No explicit rigorous realizations of such operator fields, or of the Hilbert space $\mathcal{H}$ of state vectors, in which they are supposed to act, have so far been possible. It is usually implicitly assumed that $\mathcal{H}$ is a Fock space of the type given by (10). However, such tensor product spaces, which lead to a rigorous construction of free quantum field theories, fail completely in the case of interacting fields.

6. Transitional period.

Further improvements in experimental methods led to the possibility of breaking up the atomic nucleus by means of collisions at very high energies. This opened up a whole world of sub-nuclear particles, interacting through new kinds of small distance forces.

Because of the general importance of particle creation and annihilation in this domain, and in view of the succes of quantum electrodynamics, attempts were made to apply the ideas of quantum field theory to these new particles and their interactions. A quantum field theory for the so called weak interactions, with neutrinos and muons as new particles, was obtained, another theory for the strong nuclear forces, involving, among other particles, the π-mesons. The overall succes was however limited, due to fundamental difficulties, such as e.g. non-renormalizability for the weak interactions. A period of stagnation followed in the sixties, in which there was a growing suspicion that quantum electrodynamics might just be an isolated case and that quantum field theory as a general framework for elementary particle physics was at a dead end.

New hopes were raised when in 1967 a unified field theory for weak and electromagnetic forces was introduced, independently, by Weinberg and Salam and in particular, when in 1971, 't Hooft showed it to be renormalizable.

The central idea of this theory comes from earlier work by Yang and Mills, who in 1954 suggested field theories in which generalized gauge tranformations played an important rôle. Gauge tranformations in quantum electrodynamics are essentially a change of fase, independently at every space-time point, in the $\psi(x)$ field. The transformations of Yang and Mills involve transformations by unitary matrices, acting on the components of a multi-component $\psi(x)$ field, again in each point differently. These transformations are called non-abelian gauge transformations.

The ideas of Yang and Mills, in the form used by Salam and Weinberg, gave new life to quantum field theory and started an explosive development. Gauge theories for the strong interactions have also been constructed, all this supported by an increasing amount of experimental knowledge and with an ever stronger tendency to further unification. This brings us to the present.

7. General quantum field theory: Non-abelian gauge theories.

The mathematical aspects of non-abelian gauge theories form the main subject of this conference. An introduction to the physical background of such theories is given in the lectures by Bais. A few general remarks may therefore by sufficient:

a. The present situation in sub-nuclear physics may be summarized as follows: It is generally believed that all phenomena of the sub-nuclear world should be described on the ultimate basis of fields. These field theories are quantum theories. In experiments one observes particles, associated with these fields in the manner indicated in 4.

b. There are still separate theories for some of the main types of particles and forces. A process of unification is going on; much of which is still quite speculative.

c. All the basic field theories (including, hopefully, the final, single unified theory) are non-abelian gauge theories, or Yang-Mills theories, which means that they are in a certain general way built around the concept of gauge transformations, as introduced by Yang and Mills in 1954.

d. Correspondingly they all have a strong flavour of differential geometry, in terms of fiber bundles and connections. This aspect does not however make a clear sense for the quantized (operator) fields, but has so far only a precise mathematical meaning for the mostly fictitious, associated "classical" fields.

e. The way experimentally relevant results can be extracted from these theories, is similar to what was indicated for quantum electrodynamics. An action integral, a formal expression, is the starting point for a formal power series, from which transition probabilities are calculated. A renormalization procedure is always necessary.

f. Related to this is the fact that a satisfactory mathematical formulation of non-abelian gauge quantum fields cannot, at present, be given. In respect to this, the situation is even worse than for quantum electrodynamics.

<u>Selected references</u>.

For the historical development of Quantum Mechanics, see: M.Jammer: The conceptual development of quantum mechanics, New York, 1966. See also: B.L.Van der Waerden (ed.): Sources of quantum mechanics, Amsterdam, 1967. This contains a collection of important early papers on quantum mechanics, together with a historical introduction by the editor. A classical exposition of the Hilbert space framework of quantum mechanics, that is still worth reading, is: J.von Neumann: Mathematische Grundlagen der Quantenmechanik, Berlin, 1932. (English translation: Princeton, 1955). For a collection of basic papers on quantum electrodynamics, starting with Dirac's paper of 1927, see: J.Schwinger (ed.): Quantum electrodynamics, New York, 1958. The best modern textbook on quantum field theory is: C.Itzykson and J.B.Zuber: Quantum field theory, New York, 1980. The authors have succeeded in reaching an acceptable compromise between the requirements of clarity and precision in mathematical expression and the need to go deep into the very complicated and often heuristic methods used in the application of quantum field theory to realistic problems in elementary particle physics. More in the usual mathematically rather naive style of elementary particle physics is: L.D.Faddeev and A.A.Slavnov: Gauge fields; introduction to quantum theory, Reading, 1980. Notwithstanding its sub-title, it is a fairly advanced book on the methods of quantum field theory, an admirable book in its kind. It stresses gauge fields and path-integral quantization and contains references to the most important papers in the development of gauge quantum fields. For an exposition of the mathematical aspects of quantum field theory, see: N.N.Bogolubov, A.A.Logunov and I.T.Todorov: Introduction to axiomatic quantum field theory, Reading, 1975. This book was written before the rise of gauge theories (the original Russian edition is from 1969). A systematic discussion of the new and formidable mathematical problems raised by the introduction of gauge fields into quantum field theory is unfortunately not available.

Instituut-Lorentz, Nieuwsteeg 18
2311 SB Leiden, the Netherlands

MONOPOLE SOLITONS

E.F.Corrigan
Department of Mathematics
University of Durham, U.K.

1. Introduction to monopoles

At a meeting on differential equations it may seem strange to begin a talk by introducing the idea of a magnetic monopole. Particularly, since magnetic monopoles have not yet appeared on the elementary particle scene as actual physical particles, like the electron, proton etc. However, the fascination of the magnetic monopole for a mathematical physicist is not only that it may actually exist and might one day be found in an experiment (although it would be nice if that were so) but also the theories constructed to describe charged particles in interaction with the electromagnetic field require substantial modifications, in unexpected directions, in order to accommodate monopoles at all.[1,2]

To give you an example of what I mean consider Maxwell's equations

$$\partial_\mu F^{\mu\nu} = j^\nu , \tag{1.1a}$$

$$\partial_\mu {}^*F^{\mu\nu} = 0, \tag{1.1b}$$

$${}^*F^{\mu\nu} = \tfrac{1}{2}\epsilon^{\mu\nu}{}_{\rho\sigma} F^{\rho\sigma} \tag{1.1c}$$

where $E_i = F^{oi}$ $i = 1,2,3$ are the components of the electric field, $H_i = \epsilon_{ijk}F^{jk}$, $i,j,k = 1,2,3$ are the components of the magnetic field and, j^μ $\mu = 0,1,2,3$ are the components of the electric current. The zero on the right hand side of equation (1.1b) indicates that there is no magnetic current or charge and, in nature, it looks as though éqns. (1.1) are asymmetric. However, the zero in equation (1.1b) is fortuitous because because it allows us to introduce a vector potential A_μ such that

$$F_{\mu\nu} = \partial_\mu A_\nu - \partial_\nu A_\mu \tag{1.2}$$

and equation (1.1b) becomes identically true whilst eqn. (1.1a) determines A_μ in terms of the electric charge and current density j^μ. Dirac[3] asked the question: Is it possible to modify the electromagnetic theory in such a way that magnetic monopoles could be included but the vector potential remains the fundamental field out of which both the electric and magnetic field are constructed? The answer is yes, but eqn. (1.2) cannot be maintained.

As an example of a simple situation we might wish to describe, consider a static monopole charge g at the origin $\underline{x} = 0$. Then there is no electric field, $\underline{E} = 0$, and $\underline{H}$ is a monopole field namely,

$$\underline{H} = \frac{g\underline{x}}{4\pi x^3} . \tag{1.3}$$

The problem is to represent $\underline{H}$ as 'closely as possible' by some vector potential $\underline{A}$ and write $\underline{H} = \underline{\nabla} \wedge \underline{A}$. As mentioned before this cannot be a vector field whose curl is the magnetic field of eqn. (1.3). However, we can construct a potential describing a semi-infinite thin bar magnet (or solenoid) - say, lying along the negative z axis with its end at $\underline{x} = 0$ - by

$$\underline{A}_{\text{magnet}} = \frac{g}{4\pi r} \frac{1 - \cos\theta}{\sin\theta} \hat{\underline{\phi}} \qquad r = |\underline{x}|, \tag{1.4}$$

where r, θ, φ are spherical polar coordinates such that,

$$\underline{H}_{\text{magnet}} = \frac{g\underline{x}}{4\pi r^3} + g\theta(-z)\delta(x)\delta(y)\hat{\underline{z}} .$$

In other words

$$\frac{g\underline{x}}{4\pi r^3} = \underline{\nabla} \wedge \underline{A}_{\text{magnet}} - g\theta(-z)\delta(x)\delta(y)\hat{\underline{z}} \tag{1.5}$$

and we see that the magnetic part of eqn. (1.2) is modified by subtracting a 'string' of flux along the negative z axis. Actually, the position or shape of the string is inconsequential since it can be moved at will by a suitably constructed gauge transformation[2]

$$\underline{A} = \underline{A}' + \underline{\nabla}\Omega .$$

It must, however, start at the position of the monopole and travel out to infinity. On the other hand, the fact that the string of flux is unobservable classically does not guarantee that it will be invisible in the quantised theory. Indeed, charged particles, such as electrons, scatter from flux tubes, (the Bohm-Aharanov[4] effect) unless the magnetic flux takes a special value. Specifically, for any electric charge q

$$\frac{qg}{4\pi} = \tfrac{1}{2} n\hbar \;^{(3)} \tag{1.6}$$

where n is some integer and, ħ is Planck's constant.

The condition (1.6) which must be satisfied for any pair of electric and magnetic charges implies that both electric and magnetic charge must always occur in multiples of a fundamental unit. This result is unusual in the unexpected way that it came about but, it is in fact true in nature, electric charge always does appear in multiples of a fundamental unit (a quark charge, or $\frac{1}{3}$ electronic charge). At the time when Dirac developed his theory of monopoles there was no other explanation of charge quantisation.

The above description of the monopole is singular since the vector potential used is itself inevitably singular. There is another description of the monopole field due to Wu and Yang[5] which makes essential use of the fibre bundle language. Briefly it works as follows. Let the monopole g sit at the origin $\underline{x} = 0$ and construct the field $\underline{H}$ at a given distance r from the monopole by the following procedure. Cover the sphere of radius r centred at the origin by at least two overlapping regions; for example, two hemispherical regions overlapping at the equator will do very well.

We may say that region ① includes the north pole and ② includes the south pole of the sphere. In region ① choose a (singular) potential $A_{①}$ with its singularity extending from the centre of the sphere to infinity crossing at the south pole. Thus $A_{①}$ is nowhere singular in region ① . Similarly define a potential $A_{②}$ with its singular line starting at the origin and crossing the sphere at the north pole. $A_{②}$ is nowhere singular in region ② . Thus, in each of the regions we can certainly write

$$\begin{aligned} \underline{H} &= \underline{\nabla} \wedge \underline{A}_{①} \quad \text{in region ①} \qquad \text{(a)} \\ \underline{H} &= \underline{\nabla} \wedge \underline{A}_{②} \quad \text{in region ②} \qquad \text{(b)} \end{aligned} \qquad (1.7)$$

and $\underline{H}$ is the monopole field (1.3). On the equator, where regions ① and ② overlap, the magnetic field is given by either expression 1.7a or b. Hence, on the equator $A_{①}$ and $A_{②}$ are related by a gauge transformation:

$$A_{①} = A_{②} + \underline{\nabla}\phi .$$

The total flux of the magnetic field out of the sphere is computable in terms of ϕ since we have

$$g = \int_{①} \underline{H}.\underline{ds} + \int_{②} \underline{H}.\underline{ds} = \int_{\text{equator}} (\underline{A}_{①}.\underline{dx} - \underline{A}_{②}.\underline{dx}) = \phi_{2\pi} - \phi_{o} . \qquad (1.8)$$

In quantum mechanics a charged particle of charge q is described by a wave function defined in the two patches; $\psi_{①}$, $\psi_{②}$ and related at the equator by

$$\psi_{①} = \exp\left(\frac{iq\phi}{\hbar} \right) \psi_{②} . \qquad (1.9)$$

Eqn. (1.9) only makes sense if $\exp(iq\phi/\hbar)$ is single valued around the equator in which case $\phi_{2\pi} - \phi_{o} = (\hbar\, 2\pi\, n)/q$. Consequently, using (1.8) we find

$$\frac{qg}{4\pi} = \tfrac{1}{2} n\hbar$$

as before.

Obviously, I have been over brief in this discussion and apologise for that. The idea was to indidcate the style of the changes having to be made to ordinary electrodynamics in order to accommodate the monopole (should it exist) but not the details.

In particular, there is an unexpected bonus because an alternative light is shed on charge quantisation. One also learns that the apparently harmless assumption that there is an extra kind of particle carrying a magnetic charge really requires new mathematics to describe it.

The current view in theoretical physics (see the lectures by Bais in this volume) is that electromagnetism is not a theory that should be considered in isolation from other sorts of interactions between elementary particles. A more complete and unifying theory of fundamental interactions must include the weak and strong interactions as well, together with an explanation for the varying strengths of the different forces, the variety of particles involved, their charges and other quantum numbers, etc., etc.[6] The context in which the unifying theory must be placed is quantum field theory and there are many models on the market differing in complexity and sophistication. Two particularly attractive ones from many points of view, including economy, are the Weinberg-Salam[7] model and the SU(5) model of Georgi-Glashow[8]. The latter provides the simplest model which actually unifies all three types of interaction, the former is the simplest model of all which is phenomenologically viable, but it is not unifying to the same extent as the SU(5) model.

All the unifying theories so far constructed are based on a non-abelian gauge theory. Thus SU(2) × U(1) is the gauge group for the Weinberg-Salam model, SU(5) for the Georgi-Glashow model. The electromagnetic field is constructed from a vector potential, eqn. (1.2), corresponding to an abelian gauge group, carefully selected as a subgroup of the unifying group (SU(2) × U(1) or SU(5)) by a spontaneous symmetry breaking mechanism - the Higgs mechanism[6], which will be elaborated below. A consequence of the Higgs mechanism is that electromagnetism is selected in such a way (at least in the unifying models such as SU(5)) that electric charges are multiples of a fundamental unit, basically because the electric charge operator is a specially chosen generator of the gauge group whose eigenvalues are each an integer multiple of a basic unit. There are no monopoles described by fundamental fields in these models since there is apparently no need to describe them in nature. However, the astonishing observation of 't Hooft and Polyakov in 1974[9] indicates that even though there are no fundamental monopoles in these theories, nevertheless, genuinely unifying theories such as SU(5) actually predict their existence as very heavy particles via a mechanism new to particle physics. The phenomenon is sufficiently novel that is cannot be accommodated satisfactorily within the framework of quantum field theory as it stands today but only inferred from classical and semiclassical calculations based on the gauge field equations[1,2]. The monopole reappears as a 'soliton' solution of the field equations and thus provides a challenge to both physicists and mathematicians. Little is known about soliton behaviour in four dimensions from either point of view, although it is tempting to carry over intuition gained from the study of two dimensional solitons (in the Korteweg de Vries or the Sine-Gordon equations for example) for which so much more is known.[10]

2. A model theory containing a monopole soliton

In this section we shall begin by describing SU(2) gauge theory[11,6] and use it to illustrate the idea of unification[12].

The ingredients of the theory are a vector potential, A_μ, and a Higg's field ϕ - the fermions are to be omitted, for simplicity. They play no role in building the monopole. Thus, we have

$$\begin{aligned} A_\mu &= -i A_\mu^a \frac{\sigma^a}{2} \qquad \mu = 0,1,2,3 \\ \phi &= -i \phi^a \frac{\sigma^a}{2} \end{aligned} \tag{2.1}$$

where there is an implied sum over $a = 1,2,3$ and σ^a are the Pauli σ-matrices $\begin{pmatrix} 0 & 1 \\ 1 & 0 \end{pmatrix}$, $\begin{pmatrix} 0 & -i \\ i & 0 \end{pmatrix}$, $\begin{pmatrix} 1 & 0 \\ 0 & -1 \end{pmatrix}$, respectively. Corresponding to the vector potential there is a field strength tensor $F_{\mu\nu}$ given by,

$$F_{\mu\nu} \equiv -i F^a_{\mu\nu}\frac{\sigma^a}{2} = \partial_\mu A_\nu - \partial_\nu A_\mu + e[A_\mu, A_\nu]_- . \tag{2.2}$$

Under a change of gauge the fields transform as follows,

$$\begin{aligned} A_\mu &= g^{-1} A'_\mu g + \frac{1}{e} g^{-1} \partial_\mu g \\ F_{\mu\nu} &= g^{-1} F'_{\mu\nu} g \\ \Phi &= g^{-1} \Phi' g . \end{aligned} \tag{2.3}$$

The covariant derivative of the field Φ is

$$D_\mu \Phi = \partial_\mu \Phi + e[A_\mu, \Phi] = -i(\partial_\mu \phi^a + e(\underline{A}_\mu \wedge \underline{\phi})^a)\frac{\sigma^a}{2} . \tag{2.4}$$

Using these ingredients the theory is defined by the Lagrangian

$$\mathscr{L} = -\frac{1}{4} F^a_{\mu\nu} F^{\mu\nu a} + (D_\mu \underline{\phi})^a (D^\mu \underline{\phi})^a - \frac{\lambda}{4}(\phi^a\phi^a - a^2)^2$$

from which may be derived the equations of motion:

$$D_\mu \underline{F}^{\mu\nu} = -e\underline{\phi} \wedge D^\nu \underline{\phi} \tag{2.5a}$$

$$D^2 \underline{\phi} = -\lambda \underline{\phi}(\underline{\phi} \cdot \underline{\phi} - a^2) \tag{2.5b}$$

$$D_\mu {}^*\underline{F}^{\mu\nu} = 0 \tag{2.5c}$$

the last being the Bianchi identity. A crucial functional of the fields A_μ, Φ is the energy, defined by

$$\mathcal{E} = \int d^3x \left\{ \tfrac{1}{2}(H_i^a H_i^a + E_i^a E_i^a) + \tfrac{1}{2} D_0\phi^a . D_0\phi^a + \tfrac{1}{2} D_i\phi^a D_i\phi^a + \frac{\lambda}{4}(\phi^a\phi^a - a^2)^2 \right\} . \tag{2.6}$$

We can see immediately the lowest energy configuration of the theory, it corresponds to

$$\underline{E} = \underline{H} = 0 \qquad |\underline{\phi}|^2 = a^2 \qquad D_\mu \underline{\phi} = 0. \tag{2.7}$$

In which case, $\underline{\phi}$ is a gauge transformation from a constant, $a\underline{n}$, where $\underline{n}$ is a unit vector. Next, we can examine small fluctuations away from this zero energy field configuration and notice that $\underline{A}_\mu . \underline{n}$, $\underline{H}.\underline{n}$, $\underline{E}.\underline{n}$ behave like an electromagnetic potential and fields, respectively, whilst the components $\underline{A}_\mu \wedge \underline{n}$ become massive with a mass $\hbar e a$, and these two components couple to the electromagnetic field and carry charges $\pm e\hbar$. The scalar field $\underline{\phi}$ loses two of its components retaining only the one in the direction $\underline{n}$. It becomes massive with a mass $\sqrt{\lambda} a\hbar$. Thus, the spectrum of the spontaneously broken theory when quantised contains a photon γ, two charged massive vector bosons of charge $\pm\hbar e$ and mass $\hbar a e$, and a neutral Higgs scalar of mass $\hbar a\sqrt{\lambda}$. The theory is not sufficiently realistic however because there is no room for a massive neutral vector boson, the Z[7]. It does, however, illustrate the way that the spontaneously broken symmetry ($\underline{\phi} \propto \underline{n}$ in the vacuum) on the one hand provides masses for vector bosons in a natural (and renormalisable) way and, on the other, picks out the electromagnetic field from the three vector potentials A_μ^a of the SU(2) gauge group. The theory does not obviously contain a monopole.

The monopole 'soliton'

Suppose we look classically for field configurations which have finite energy and which satisfy the field equations (2.5). Following 't Hooft and Polyakov(9,1,2) we first examine the behaviour of the fields asymptotically. Clearly, for finite energy the boundary conditions will have to be:

$$|\underline{\phi}|^2 \longrightarrow a^2 \tag{2.8}$$

$$D_\mu \underline{\phi} \longrightarrow 0 \tag{2.9}$$

$$E_i^a, H_i^a \longrightarrow 0. \tag{2.10}$$

If we examine (2.9) closely we can use it to constrain the vector potential asymptotically. Specifically

$$\underline{A}_\mu = \frac{1}{e}\frac{\partial_\mu \underline{\phi} \wedge \underline{\phi}}{a^2} + \frac{1}{a}\underline{\phi}\,\alpha_\mu \tag{2.11}$$

with a corresponding field strength

$$\underline{F}_{\mu\nu} = \frac{\underline{\phi}}{a}(\partial_\mu \alpha_\nu - \partial_\nu \alpha_\mu - \frac{1}{ea^3}\underline{\phi}.\partial_\mu \underline{\phi} \wedge \partial_\nu \underline{\phi}). \tag{2.12}$$

Notice that the field strength is parallel to $\underline{\phi}$ and that $\underline{\phi}.\underline{F}_{\mu\nu}/a$ is gauge invariant. Interesting solutions are obtained by supposing $\underline{\phi}$ is independent of x^0, the time, working in the gauge $\underline{A}^0 = 0$ and setting $\alpha_\mu \equiv 0$. Then $\underline{E} = 0$, whilst for the magnetic part of the field strength,

$$H_k = -\frac{1}{2ea^3}\,\underline{\phi}\,.\,\partial_i \underline{\phi} \wedge \partial_j \underline{\phi}\;\varepsilon_{ijk}\,. \tag{2.13}$$

Of course the field $\underline{\phi}$ that enters eqn. (2.13) need only satisfy (2.8). In particular, a possibility is

$$\underline{\phi} \sim a\hat{\underline{x}}\,, \tag{2.14}$$

for which the asymptotic magnetic field is easily computed to be

$$\underline{H} \sim -\frac{1}{e}\frac{\underline{x}}{|\underline{x}|^3} \tag{2.15}$$

- a magnetic monopole field!

't Hooft and Polyakov argued that there is a finite energy solution to the equations of motion (2.5) with the boundary condition (2.14) for which the fields A_i^a, ϕ^a are regular everywhere. In other words, the solution describes a monopole without the necessity of any source. A detailed look at the energy density shows that it is strongly localised near the origin - which may be considered the position of the monopole, particularly as the scalar ϕ vanishes there. The total energy turns out to be proportional to a/e times a function of λ. In terms of the masses and couplings of the theory the energy of the solution is of the order

$$\frac{M_{\text{vector boson}}}{e^2} .$$

If e is the usual electronic charge unit and M is the popular estimate of a vector boson mass (in the region of 80 GeV) then the energy of this solution, which we can think of as the mass of a static monopole, is around 10^4 GeV[2]. This is very heavy on the scale of the known particles.

The topological nature of the monopole is underlined by the following observation[13,1,2]. Since $|\underline{\phi}|^2 \to a^2$ asymptotically we may think of the Higgs field as providing a mapping $S_2 \to S_2$ at ∞. The S_2 at $|\underline{x}| \to \infty$ is mapped to the orbit of $\underline{\phi}$ itself which is also a spherical surface of radius a. Such maps fall into homotopy classes, members of each of which can be <u>continuously</u> deformed into one another and, conveniently labelled by the integers $\mathbb{Z}$. It turns out that the integral defining the magnetic flux of the field (2.13) is

$$G = \int \underline{dS} . \underline{H} = \frac{1}{2e} \int dS_k \frac{\underline{\phi} . \partial_i \underline{\phi} \wedge \partial_j \underline{\phi}}{a^3} \varepsilon_{ijk} \tag{2.16}$$

and is just the analytic way of computing the integers associated with the class to which a particular $\underline{\phi}$ belongs.[14] It is the topological nature of the boundary value of $\underline{\phi}$ at ∞ that provides the stability of the monopole. (In much the same way as the boundary conditions, $u(\infty) \to 2k_\infty \pi$, $u(-\infty) \to 2k_{-\infty}\pi$, $k_\infty, k_{-\infty}$ integers, lead to the stability of a multisoliton solution u(x,t) of the Sine-Gordon equation:

$$u_{tt} - u_{xx} = -\sin u ,$$

and corresponds to a multisoliton of 'charge' $(k_\infty - k_{-\infty})$[10]).

Clearly, once the monopole has been spotted there are many questions to be answered. Ultimately we should like to know what kind of a role such a curious classical solution will play in quantum field theory. That is far from clear since there is no formalism within which a particle-like solution like the monopole fits naturally. It is not known to be described by a local field - though that may indeed be the case if something like the curious duality illustrated by the Sine-Gordon - Thirring theory[15] also occurs in the four-dimensional gauge theory. Despite a lack of formal understanding of the monopole, physcists have speculated recently about their cosmological significance during the early moments of the universe.[16]

Quite apart from its physical interest, the monopole is mathematically exciting because it provides a candidate for a non-trivial example of a four-dimensional soliton. There are several questions to ask. The monopole discussed by 't Hooft and Polyakov is a spherically symmetric solution (in the sense that a spatial rotation may be compensated by a gauge transformation[1,2]) and corresponds to the smallest non-zero 'topological charge'. A natural question to ask is are there more complicated solutions corresponding to multiple charges interacting with each other - in rather the same way as the Sine-Gordon theory or KdV theory have multisoliton solutions?[17] Is there any mathematical structure underlying the monopoles as there is for the instantons - (solutions to $F_{\mu\nu} = {}^*F_{\mu\nu}$ with finite action[18])? The remaining parts of these lectures will be devoted to recent progress on the latter question.

3. The BPS monopole[19,20,21]

If we start the task of solving eqns. (2.5) by seeking a solution as a perturbation series in λ and concentrate, in the first instance, on the term of order λ^o it is possible to write down an analytic representation of the monopole. It is[20]

$$\phi_a = \frac{x_a}{r}\left(\frac{1}{r} - \coth r\right) = -\partial_a \ln\frac{\mathrm{sh}\, r}{r}, \qquad r = |\underline{x}| \tag{3.1}$$

$$A_i^a = \varepsilon_{aij}\frac{x_j}{r^2}\left(1 - \frac{r}{\mathrm{sh}\, r}\right) \tag{3.2}$$

$$A_o^a = 0.$$

This solution has the spherical symmetry alluded to earlier, the boundary conditions corresponding to unit magnetic charge (in units for which $h = e = a = 1$, for convenience) and is everywhere regular.

The solution (3.1) has an additional important property, the SU(2) 'magnetic' field and the covariant derivative of the Higgs field ϕ are simply related:

$$H_i^a = (D_i\underline{\phi})^a. \tag{3.3}$$

Equation (3.3) (which implies the field equations (2.5) for $\lambda = 0$)[19] is of crucial importance in what follows.

Reconsider the energy functional, 'completing the square':

$$\mathcal{E} = \int d^3x \left\{ \tfrac{1}{2}\underline{E}^a.\underline{E}^a + \tfrac{1}{2}(D^o\underline{\phi})^a(D^o\underline{\phi})^a + \tfrac{1}{2}(H_i^a - (D_i\underline{\phi})^a)(H_i^a - (D_i\underline{\phi})^a) + \frac{\lambda}{4}(\underline{\phi}\cdot\underline{\phi} - a^2)^2 \right\} + \int d^3x\, H_i^a (D_i\phi)^a$$

$$\geqslant \int d^3x\, H_i^a D_i \phi^a = \int d^3x\, \nabla_i (H_i^a\phi^a) = 4\pi G. \tag{3.4}$$

We notice that the energy functional, for field configurations having a boundary condition appropriate to the homotopy class N of $\underline{\phi}$, is bounded below by $4\pi N$ (in our special units) and the bound is saturated if and only if:

$$E^a = 0, \qquad (D^o\underline{\phi})^a = 0, \qquad \lambda = 0, \tag{3.5}$$

and eqn. (3.3) holds. The solution (3.1) satisfies all these and saturates the bound for $N = 1$. (The bound for $|N| = 1$, really, since a solution of charge -1 is obtained from (3.1 by sending $\underline{x} \to -\underline{x}$.)

We can also remark that (3.3) and (3.5) imply, via (3.4), that the energy density $\varepsilon(x)$ is given entirely by $\underline{\phi}$:

$$\varepsilon(x) = \tfrac{1}{2}\nabla^2 |\underline{\phi}|^2 , \qquad (3.6)$$

an elegant formula.

We finally note from eqn. (3.1),

$$|\underline{\phi}|^2 = 1 - \nabla^2 \ell n \frac{\text{shr}}{r} . \qquad (3.7)$$

From a more physical point of view the limit $\lambda = 0$ is intriguing. In that limit the Higgs particle in the spontaneously broken gauge theory becomes massless (its mass was proportional to $\sqrt{\lambda}$) and corresponds to a long range attractive force (since it is a scalar field). Manton[21] pointed out that two widely separated monopoles of like charge, which you might expect to experience a long range coulomb force because of their magnetic charge, actually experience no force at all - the (repulsive) coulomb force is cancelled by the Higgs force. More recently, Taubes[22] proved the astonishing result that for any given magnetic charge N the equations (3.3), (3.4) have a regular solution, with the correct boundary conditions on $\underline{\phi}$, and the total energy of the solution $4\pi N$ - just N times the energy of the single monopole. In other words, in the limit $\lambda = 0$ the multimonopoles are static and 'non-interacting' - very unlike the situation in the Sine-Gordon theory where multiple soliton solutions have to be time dependent, and interacting.[17]

This result suggests that it ought to be possible to understand the nature of the solutions to eqn. (3.3).

One final word for this section. It is possible to count the number of parameters, or degrees of freedom, that a multimonopole solution to eqn. (3.3) should have. It is $4N-1$ for a solution of total charge N.[23] 3N of these parameters may be thought of as the positions in 3 space of the N monopoles though the other $N-1$ degrees of freedom are mysterious. It is likely, however, that any explicit solution will contain a number of parameters and that some combinations of them will be interpretable as the monopole positions (e.g. the zeros of the Higgs field $\underline{\phi}$?) The question is how to find them.

4. Multimonopoles

At first sight one might imagine that monopoles placed along a line would be the next simplest case to consider because it might be expected to have an axial symmetry. This is not so, however.[24] The axisymmetric case has been investigated thoroughly by Forgacs, Horvath and Palla[25] and they discovered that whilst there are axially symmetric solutions with multiple magnetic charge they do not correspond to separated monopoles. Instead, the energy density of the multimonopole is concentrated on a ring surrounding the origin in a plane containing the origin whose diameter is roughly proportional to the monopole strength. (Unlike the spherically symmetric single monopole whose energy is concentrated about the origin, or, more generally about the zero of the Higgs field). The 'axis' of the axisymmetric solutions is the normal through the origin to the plane containing the energy density. All these solutions, whatever the magnetic charge, contain just five parameters (the origin together with the direction of the axis of symmetry) and constitute a small subset of the total solution space. Nevertheless, the way in which they have been obtained is very interesting.

An axisymmetric ansatz for eqn. (3.3)[26] may be written

$$\begin{aligned} \underline{\phi} &= \frac{1}{f}\frac{\partial}{\partial z}(0,g,-f) \\ \underline{A}_\psi &= \frac{1}{f}\frac{\partial}{\partial \rho}(0,g,-f) \\ \underline{A}_z &= \frac{1}{f}\frac{\partial}{\partial z}(g,0,0) \\ \underline{A}_\rho &= \frac{1}{f}\frac{\partial}{\partial \rho}(g,0,0) \end{aligned} \tag{4.1}$$

where ρ, z, ψ are cylindrical polar coordinates and the two functions f,g are functions of (ρ,z) only. Putting f and g together,

$$\varepsilon = f + ig \tag{4.2}$$

the quantity ε satisfies Ernst's equation,[27]

$$\mathrm{Re}\,\varepsilon\,\nabla^2\varepsilon - (\nabla\varepsilon)^2 = 0, \tag{4.3}$$

familiar to relativists in the context of axisymmetric solutions to Einstein's field equations.

The Ernst equation has been extensively investigated and is known to have several Bäcklund transformations[10,28] associated with it. In particular, Forgacs, Horvath and Palla discovered a combination of Bäcklund transformations which maintained $|\underline{\phi}|^2 = 1$ asymptotically, (in units for which a = 1) but changed the magnetic charge by shifting the field ϕ from one homotopy class at infinity to the neighbouring one. Usually, Bäcklund transformations introduce free parameters (cf. the velocity parameters in the Sine-Gordon multi-soliton solutions[17]) but, in this case, the requirement that the solution should be everywhere regular restricts the extra parameters to special values-leaving only the five free parameters mentioned before.

The relationship of the axisymmetric monopole with Bäcklund transformations reinforce the idea that the monopole is a soliton.

It is worth remarking, whilst we are talking about symmetric solutions, that it is possible to investigate spherically symmetric monopoles in larger gauge groups. Then one finds that the variety of monopoles is greater but for some of them an appropriate choice of ansatz for the Higgs field and the vector potential is given in terms of functions satisfying the Toda lattice equations, with an exponential interaction governed by the Cartan matrix of the gauge group. Work on this has been extensively reviewed.[29] Again it tends to support the idea that the monopole is a soliton in four dimensions. Of course, whilst the special solutions known are tending to indicate the soliton nature of the monopole the acid test is to find time-dependent scattering solutions and to examine them in detail. At the moment progress in that direction looks hard.

The axisymmetric solutions may also be obtained another way. Indeed, there is nothing in what follows that requires any symmetry to be imposed apriori and the formal structure is broad enough to permit the description of solutions with 4N - 1 parameters.

The first step is to understand eqn. (3.3) from a different point of view. If we work in the gauge $A_0^a = 0$ then Taubes' theorem guarantees a solution which is independent of x^0. Introduce a Euclidean variable x^4 on which nothing depends and set

$$A_4^a = \phi^a \tag{4.4}$$

$$F_{i4}^a = (\partial_i A_4 - \cancel{\partial_4 A_i} + [A_i, A_4])^a = (D_i\phi)^a .$$

Then eqn. (3.3) reads

$$F_{\mu\nu} = {}^*F_{\mu\nu} \equiv \tfrac{1}{2}\varepsilon_{\mu\nu}{}^{\rho\sigma} F_{\rho\sigma} \qquad \mu,\nu = 1,2,3,4 \tag{4.5}$$

which are the equations of self duality, much studied recently[18]. However, unlike the instanton case we are here interested in solutions for which A_μ, $\mu = 1,2,3,4$ is independent of x^4, $\underline{A}_4 . \underline{A}_4 \to 1$ as $|\underline{x}| \to \infty$, and the total energy (rather than action) is finite. Nevertheless, the machinery set up to study the instantons, suitably adapted, is useful here also.

The Yang equations[30]

One way to proceed, faced with eqn. (4.5), is to complexify the four dimensional Euclidean space and introduce new variables as follows,

$$x = x_4 + i\underline{x}.\underline{\sigma} = \begin{pmatrix} x_4 + ix_3 & x_2 + ix_1 \\ -x_2 + ix_1 & x_4 - ix_3 \end{pmatrix} = \begin{pmatrix} y & -\bar{z} \\ z & \bar{y} \end{pmatrix} . \tag{4.6}$$

(Note: the bar denotes complex conjugation (*) only when x_μ is real). The utility of these variables is apparent when eqn. (4.5) is expressed in terms of them. We get

$$F_{yz} = F_{\bar{y}\bar{z}} = 0, \qquad F_{y\bar{y}} + F_{z\bar{z}} = 0 \tag{4.7}$$

from which we deduce

$$\begin{pmatrix} A_y \\ A_z \end{pmatrix} = D^{-1}\begin{pmatrix} \partial_y \\ \partial_z \end{pmatrix} D \quad \text{and} \quad \begin{pmatrix} A_{\bar{y}} \\ A_{\bar{z}} \end{pmatrix} = \bar{D}^{-1}\begin{pmatrix} \partial_{\bar{y}} \\ \partial_{\bar{z}} \end{pmatrix} \bar{D} \tag{4.8}$$

where $D^+ = \bar{D}^{-1}$ when x_μ is real and, D, $\bar{D}$ are elements of $SL(2,C)$.

The combination $J = D\bar{D}^{-1}$ is gauge invariant, and conveniently parametrised by three functions ε, ϕ, γ:

$$J = D\bar{D}^{-1} = \frac{1}{\phi}\begin{pmatrix} 1 & \gamma \\ -\varepsilon & \phi^2 - \varepsilon\gamma \end{pmatrix} = \underbrace{\frac{1}{\sqrt{\phi}}\begin{pmatrix} 1 & 0 \\ -\varepsilon & \phi \end{pmatrix}}_{D}\underbrace{\frac{1}{\sqrt{\phi}}\begin{pmatrix} 1 & \gamma \\ 0 & \phi \end{pmatrix}}_{\bar{D}^{-1}} \tag{4.9}$$

The splitting indicated in the final step of eqn. (4.9) implies a gauge choice. Eqns. (4.8) imply the first pair of equations (4.7) whilst the last (in terms of ε, ϕ, γ) reads,

$$\partial^2 \ln\phi = \frac{4}{\phi^2}(\partial_y \varepsilon\, \partial_{\bar{y}}\gamma + \partial_z \varepsilon\, \partial_{\bar{z}}\gamma)$$

$$\left(\frac{\partial_y \varepsilon}{\phi^2}\right)_{\bar{y}} + \left(\frac{\partial_z \varepsilon}{\phi^2}\right)_{\bar{z}} = 0 \tag{4.10}$$

$$\left(\frac{\partial_{\bar{y}}\gamma}{\phi^2}\right)_y + \left(\frac{\partial_{\bar{z}}\gamma}{\phi^2}\right)_z = 0,$$

and constitutes a set of three coupled equations for ϕ, ε, and γ.

The equations (4.10) are interesting in their own right and have a couple of useful properties which will be referred to as Bäcklund transformations.[31]

If ε, ϕ, γ are solutions to (4.10) so also are e, f, g defined by

(A) $$\phi = \frac{1}{f}, \quad \begin{pmatrix} \partial_y \\ \partial_z \end{pmatrix}\varepsilon = -\frac{1}{f^2}\begin{pmatrix} -\partial_{\bar{z}} \\ \partial_{\bar{y}} \end{pmatrix} g, \quad \begin{pmatrix} -\partial_{\bar{z}} \\ \partial_{\bar{y}} \end{pmatrix}\gamma = -\frac{1}{f^2}\begin{pmatrix} \partial_y \\ \partial_z \end{pmatrix} e \tag{4.11}$$

(B) $$\begin{pmatrix} e & f \\ f & g \end{pmatrix} = \begin{pmatrix} \varepsilon & \phi \\ \phi & \gamma \end{pmatrix}^{-1} \tag{4.12}$$

The latter constitutes a gauge transformation on the vector potentials defined by eqn. (4.8) with gauge function

$$g = \frac{\begin{pmatrix} \gamma & \phi \\ \phi & \varepsilon \end{pmatrix}}{\sqrt{\phi^2 - \varepsilon\gamma}}\,. \tag{4.13}$$

An example of the use of (A) and (B) is to observe that they recursively linearise equns. (4.10). In other words, if we write $\Phi_\ell = (\varepsilon_\ell, \phi_\ell, \gamma_\ell)$ then the definition

$$\Phi_\ell = AB\Phi_{\ell-1}\,, \tag{4.14}$$

yields a convenient expression for ε_ℓ, ϕ_ℓ, γ_ℓ in terms of a set of functions Δ_r in the following way.[31]

$$\begin{pmatrix} \partial_y \\ \partial_z \end{pmatrix}\Delta_r = \begin{pmatrix} -\partial_{\bar{z}} \\ \partial_{\bar{y}} \end{pmatrix}\Delta_{r+1} \qquad (\Longrightarrow \partial^2\Delta_r = 0) \tag{4.15}$$

and

$$\begin{pmatrix} \varepsilon_\ell & \phi_\ell \\ \phi_\ell & \gamma_\ell \end{pmatrix}^{-1} = \text{corner entries of} \begin{pmatrix} \Delta_{-\ell} & \Delta_{-\ell+1} & \cdots\cdots & \Delta_{-1} & \Delta_0 \\ \Delta_{-\ell+1} & & \Delta_{-1} & \Delta_0 & \Delta_1 \\ \vdots & & & & \vdots \\ \Delta_{-1} & & & & \vdots \\ \Delta_0 & \Delta_1 & \cdots\cdots & \Delta_{\ell-1} & \Delta_\ell \end{pmatrix}^{-1} . \tag{4.16}$$

Eqn. (4.16) follows from eqn. (4.14) iteratively with the additional definition

$$\Phi_0 = (\Delta_0, \Delta_0, \Delta_0), \qquad \partial^2 \Delta_0 = 0, \qquad B\Phi_0 = \left(\frac{1}{\Delta_0}, \frac{1}{\Delta_0}, \frac{1}{\Delta_0} \right) \tag{4.17}$$

to start the iteration going. More details of that can be found in ref. (31). At this stage it is worth remembering that we are eventually seeking solutions to equations (4.7) independent of x_4, strongly suggesting that each of the functions Δ_r should have the same x_4 dependence, so that it will cancel out in eqn. (4.8). Then, since z, $\bar{z}$ do not depend on x_4 whereas y and $\bar{y}$ do, eqn. (4.15) implies

$$\Delta_r = f(x_4)\tilde{\Delta}_r(x_1,x_2,x_3) \qquad \forall r; \quad \frac{\partial_4 f}{f} = \text{constant}, \tag{4.18}$$

i.e.

$$f = \exp(ikx_4), \quad \text{and} \quad \nabla^2 \tilde{\Delta}_r = k^2 \tilde{\Delta}_r . \tag{4.19}$$

We note that the gauge transformation (4.13) is also x_4 independent with this special dependence of the Δ's on x_4. We may note further that $\underline{A}_4$ may be computed directly from eqns. (4.16) and (4.8) and we can examine the large $|\underline{x}|$ behaviour of $|\underline{A}_4|^2$. Following Prasad[32], (a cunning use of A,B above)[33], we find

$$|\underline{A}_4|^2 = |\underline{\phi}|^2 = k^2 - \nabla^2 \ln \det \tilde{D}_\ell , \tag{4.20}$$

where

$$(\tilde{D}_\ell)_{ij} = \tilde{\Delta}_{i+j-\ell-1} \qquad i,j = 1,\ldots,\ell . \tag{4.21}$$

(The matrix appearing in eqn. (4.16) in $D_{\ell+1}$ in the same notation.)

Eqn. (4.20) is worthy of comment for several reasons,

(i) We see that $k = 1$ to agree with $|\phi|^2 \to 1$ asymptotically.

(ii) Eqn. (4.20) is a generalisation of eqn. (3.7). Indeed, if we set $\Delta_0 = \frac{\text{shr}}{r}\exp(ix_4)$, we see that $\tilde{\Delta}_0$ satisfies $\nabla^2\tilde{\Delta}_0 = \tilde{\Delta}_0$ as it should.

(iii) Each of the Δ's satisfies the Helmholtz equation. So, for large $|\underline{x}|$ we expect $\tilde{\Delta}_k \sim (e^r/r)c_k$ for each $k = 0, \pm 1, \ldots$

Then

$$|\underline{\phi}|^2 \sim 1 - \frac{2\ell}{r} \qquad \text{as } r \to \infty . \tag{4.22}$$

In other words, eqn. (4.20) automatically leads to the correct boundary condition for a monopole of charge ℓ! (cf. eqn. (3.6) and compute the total energy - it is given by the coefficient of $1/r$ in the asymptotic expansion of $|\underline{\phi}|^2$.)

(iv) We are tempted to regard (4.20) as a nonlinear superposition principle for monopoles similar to the multisoliton formulae (of Hirota[34], for example) which put together solutions to linear equations in a non-linear way.[17] The similarity is very striking.

The next problem we face is to choose the Δ's so that eqns. (4.15), (4.18), (4.19) are satisfied, their asymptotic behaviour is given by (iii), above, and the vector potentials, A_μ^a, are everywhere regular and real.

The twistor formalism[35,36,37]

There is an alternative approach to the self-dual equations, developed by Atiyah and Ward (see Richard Ward's lectures this volume), which supplies additional information about the functions Δ_r.

In the four-dimensional complex Euclidean space introduced at the start of the last section we may note that there are completely null planes, that is, 2-dimensional subspaces spanned by pairs of mutually orthogonal null vectors, k_μ, h_μ. Also, the null planes come in two types, self-dual and anti-dual, in the following sense:

$$k_\mu h_\nu - k_\nu h_\mu = \pm\tfrac{1}{2}\,\varepsilon_{\mu\nu}{}^{\rho\sigma}\,(k_\rho h_\sigma - k_\sigma h_\rho). \tag{4.23}$$

Furthermore, the points of an anti-dual plane may be specified by,

$$x\pi = \omega \tag{4.24}$$

where x is given by eqn. (4.6) and, (π,ω) are complex 2-vectors characterising the null plane. Any non-zero complex multiple of (π,ω) specifies the same plane and we may regard the set of all planes as three dimensional complex projective space CP^3 (minus a line).

A self-dual field strength $F_{\mu\nu}$ is trivial on an antidual plane and this fact was used by Ward to set up an alternative description of self-dual fields in terms of two dimensional holomorphic vector bundles over CP^3 (minus a line). On the one hand the integrability of the gauge theory connection on the null planes is used to set up the vector bundle and on the other, and more importantly for our purposes, the procedure is reversible and from any holomorphic vector bundle over CP^3 we can recover a self-dual gauge potential, defined on the real four dimensional Euclidean space we started with.

To describe a vector bundle over CP^3 (minus a line) we need two coordinate patches ① and ② and a matrix that 'patches' together the two pieces. A convenient choice of coordinates for CP^3 are $x_{①}$ and $x_{②}$,

$$\underbrace{\begin{pmatrix} 0 & \omega_1/\pi_2 \\ 0 & \omega_2/\pi_2 \end{pmatrix}}_{x_{①}}\begin{pmatrix}\pi_1\\ \pi_2\end{pmatrix} = \begin{pmatrix}\omega_1\\ \omega_2\end{pmatrix},\qquad \underbrace{\begin{pmatrix} \omega_1/\pi_1 & 0 \\ \omega_2/\pi_1 & 0 \end{pmatrix}}_{x_{②}}\begin{pmatrix}\pi_1\\ \pi_2\end{pmatrix} = \begin{pmatrix}\omega_1\\ \omega_2\end{pmatrix} \tag{4.25}$$

on the patches $\pi_2 \neq 0$, $\pi_1 \neq 0$, respectively. For convenience we define the complex variables μ, ν and ξ by:

$$\pi_1/\pi_2 = \xi, \quad i\omega_2/\pi_2 = \mu = ix_4 + x_3 - (x_1 + ix_2)\xi,$$
$$i\omega_1/\pi_1 = \nu = ix_4 - x_3 - (x_1 - ix_2)\frac{1}{\xi}. \tag{4.26}$$

The patching matrix is defined in the overlap of ① and ② (i.e. almost everywhere) and is a function of μ, ν and ξ. In terms of ξ, patch ① excludes $\xi = \infty$, patch ② excludes $\xi = 0$. In the overlap region $x_{①}$ and $x_{②}$ label the same null plane and the patching matrix is

$$g_{21} = P \exp \int_{x_{②}}^{x_{①}} A_\mu \, dx_\mu \tag{4.27}$$

and corresponds to integrated parallel transport in the null plane from $x_{①}$ to $x_{②}$. (The P denotes path ordering but in fact g is independent of the path taken because $F_{\mu\nu}$ is trivial on the anti-dual plane). Thus, given a vector potential g_{21} is constructed via eqn. (4.27). On the other hand, if we are given g_{21} we can determine the vector potential at any point of the null plane by writing

$$g_{21} = P \exp \int_{x_2}^{x} A_\mu \, dx_\mu \quad P \exp \int_{x}^{x_1} A_\mu \, dx_\mu \tag{4.28}$$

taking a path in the null plane via the point we are interested in. Note that on the null plane the path ordered integrals only depend on their end points and not on the path taken. That is,

$$g_{21} = h(x_1,x) \quad k^{-1}(x_2,x) \tag{4.29}$$

where h is regular at $\xi = \infty$, k regular at $\xi = 0$. Then we have

$$\text{(i)} \qquad \left[\begin{pmatrix} \partial_y \\ \partial_z \end{pmatrix} - \xi \begin{pmatrix} -\partial_{\bar{z}} \\ \partial_{\bar{y}} \end{pmatrix} \right] g = 0 \tag{4.30}$$

as an identity, and

$$\text{(ii)} \qquad \begin{pmatrix} A_y \\ A_z \end{pmatrix} - \xi \begin{pmatrix} -A_{\bar{z}} \\ A_{\bar{y}} \end{pmatrix} = h^{-1} \left[\begin{pmatrix} \partial_y \\ \partial_z \end{pmatrix} - \xi \begin{pmatrix} -\partial_{\bar{z}} \\ \partial_{\bar{y}} \end{pmatrix} \right] h$$
$$= k^{-1} \left[\begin{pmatrix} \partial_y \\ \partial_z \end{pmatrix} - \xi \begin{pmatrix} -\partial_{\bar{z}} \\ \partial_{\bar{y}} \end{pmatrix} \right] k. \tag{4.31}$$

Thus, given g_{21} we can recover D, $\bar{D}$ (eqn. 4.8) as

$$D = k(x,\xi = 0), \qquad \bar{D} = h(x,\xi = \infty). \tag{4.32}$$

The monopole problem is now phrased a different way - which matrices g_{21} give rise to the sort of solution we want to find? We note,[38,39]

(1) There is a vast redundancy in g_{21} since, any two related via a 'gauge' transformation,

$$g_{21} = a\, g'_{21} A, \tag{4.33}$$

where a and A are SL(2,C) matrix functions of μ, ν, ξ regular at $\xi = \infty$, $\xi = 0$ respectively, give rise to gauge equivalent vector potentials, (in the sense of eqn. (2.3)).

(2) A_μ is to be independent of x_4 so there must be a gauge in which g_{21} is independent of x_4. This is a stronger condition than it looks at first sight.

(3) A_μ is to be real when x belongs to the real Euclidean space we started with. This means that there must be a gauge in which

$$g_{21}(\xi) = [g_{21}(-1/\xi^*)]^+, \tag{4.34}$$

following directly from eqn. (2.31). This is also a strong condition.(39)

(4) g_{21} has to be constructed in such a way that the total energy of the monopole solution is finite automatically.

Atiyah and Ward(37) argued that for instantons it was enough to take

$$g_{21} = \begin{pmatrix} \xi^\ell & \rho(\mu,\nu,\xi) \\ 0 & \xi^{-\ell} \end{pmatrix} \qquad \ell = 0,1,2,\ldots \tag{4.35}$$

Writing

$$\rho(\mu,\nu,\xi) = \sum_{-\infty}^{\infty} \xi^r \Delta_{-r} \tag{4.36}$$

we note that eqn. (4.30) automatically implies eqn. (4.15) so that ρ is a sort of generating function for the Δ's we had before. Furthermore,(31) computing D, $\bar{D}$ via eqns. (4.32) we recover precisely the expression for the vector potentials implied by eqn. (4.16) and eqn. (4.8). Let us assume that this form of g_{21} is sufficient in the monopole case also.

We argued previously that the functions Δ_r ought to have the form

$$\Delta_r = \exp(ix_4)\tilde{\Delta}_r \tag{4.37}$$

so that we expect, from eqn. (4.36), $\rho = \exp(ix_4)\tilde{\rho}$. However, this cannot be since ρ is a function of the combinations μ, ν and ξ. We can say, however, that ρ has the form

$$\rho = \exp\left(\frac{\mu+\nu}{2}\right)\tilde{\rho}\left(\frac{\mu-\nu}{2}, \xi\right). \tag{4.38}$$

Setting $\gamma = \frac{\mu-\nu}{2}$, it is easy, once shown how,(38) to find matrices a, A such that g_{21} is gauge equivalent to

$$\begin{pmatrix} \xi^\ell e^\gamma & \tilde{\rho}(\gamma,\xi) \\ 0 & \xi^{-\ell} e^{-\gamma} \end{pmatrix} \tag{4.39}$$

which is independent of x_4 altogether. Condition (2) has effectively reduced ρ to a function of two variables only, γ and ξ.

Ward(38) pointed out that if $\ell = 1$, taking $\tilde{\rho} = \frac{\mathrm{sh}\gamma}{\gamma}$ leads to $\Delta_o = \exp(ix_4)\frac{\mathrm{sh}r}{r}$, and hence to the B.P.S. monopole. The choice $\ell = 2$ and $\tilde{\rho} = \text{ch}\gamma/(\gamma^2 + \frac{\pi^2}{4})$ (38) leads to the cylindrically symmetric monopole of Forgacs et al. The choice $\ell = 3$ and $\tilde{\rho} = \text{sh}\gamma/\gamma(\gamma^2 + \pi^2)$ leads to a cylindrically symmetric monopole of charge 3, etc.(40) We may note that the functions $\tilde{\rho}$ are designed to be singularity-free but, regularity mysteriously constrains the denominators. For example, $\tilde{\rho} = \text{ch}\gamma/(\gamma^2 + \frac{9\pi^2}{4})$ will not work.(38) The latter observation means that condition (4) will be difficult to imlement, but it is at least necessary that:

(5) the functions $\tilde{\rho}$ have no singularities in the ξ plane that move with the spatial coordinates x_i, $i = 1,2,3$.

Condition (3) has been analysed fully.(41,39) In order to satisfy it and the condition (5) g_{21} has to be gauge equivalent to a matrix containing precisely $4\ell - 1$ real parameters.(39) It is gratifying that this is the number of degrees of freedom of a monopole of charge ℓ. It is not, however, clear that the solutions so obtained are everywhere regular.

5. Conclusion and outlook

I have attempted to convince you that the investigation of monopoles is an interesting exercise both from the point of view of physics and of mathematics. Regrettably, the arguments have to be somewhat sketchy and incomplete but perhaps their flavour and depth have become apparent, if not all the details. There is still much to be done. All the detailed work so far has involved a study of the equation (3.3) and its counterpart for bigger groups and therefore static monopoles. We still do not know if the monopole is really a soliton because we do not yet know any systematic way to discover time dependent solutions to the full field equations. Nevertheless, there is an astonishing structure even in the limited cases that have been studied and it is not yet fully elaborated. It would be interesting to know for example, the precise relationship between the Bäcklund transformation view of the cylindrically symmetric monopoles and their description in the Atiyah-Ward formalism. The regularity of the general solutions of ref. (39) has yet to be proved although recent work of Nahm may be helpful in that respect.(42)

Given that the classical solutions will eventually be understood there remains the question of the role of the monopole in quantum field theory.

References

For reviews, see

1) S. Coleman, New Phenomena in Subnuclear Physics. Ed. A. Zichichi (New York, Plenum) 1975, 297.
2) P. Goddard and D. Olive, Reports on Progress in Physics 41 (1978) 1357.

Suggestions as to the meaning of the monopole occurred in

P. Goddard, J. Nuyts and D. Olive, Nucl. Physics B125 (1979) 1.
C. Montonen and D. Olive, Phys. Lett. B728 (1977) 117.

3) P.A.M.Dirac, Proc. Roy. Soc. A133 (1931) 60.
Phys. Rev. 74 (1948) 817.
4) Y. Aharonov and D. Bohm, Phys. Rev. 123 (1961) 1511.
5) T.J.Wu and C.N.Yang, Phys. Rev. D12 (1975) 3845.
6) See for example J.C.Taylor, Gauge Theories of Weak Interactions (Cambridge University Press) 1976.
7) A. Salam, Proc. 8th Nobel Symposium: Elementary particle theory. Ed. N. Svartholm (New York, Wiley) 1968, 337.
S. Weinberg, Phys. Rev. Lett. 19 (1967) 1264.
8) H. Georgi and S.L.Glashow, Phys. Rev. Lett. 32 (1974) 438.
9) G. 't Hooft, Nucl. Phys. B79(1974) 276.
A.M.Polyakov, JETP Lett. 20 (1974) 194.
10) See, for example
G.L.Lamb, Elements of Soliton Theory (New York, Wiley) 1980.
H.B.Thacker, Rev. Mod. Phys. 53 (1981) 253.
11) R.L.Mills and C.N.Yang, Phys. Rev. 96 (1954) 191.
12) H. Georgi and S.L.Glashow, Phys. Rev. D6 (1972) 2977.
13) Yn. S. Tyupkin, V.A.Fateev and A.S.Shwarz, JETP Lett. 21 (1975) 41
M.I.Monastyrskii and A.M.Perelomov, JETP Lett. 21(1975) 43.
14) J. Arafune, P.G.O.Freund and C.J.Goebel, J. Math. Phys. 16 (1975) 433.
15) S. Coleman, Phys. Rev. D11 (1975) 2088.
S. Mandelstam, Phys. Rev. D11(1975) 3026.
16) See, for example
J. Ellis, M.K.Gaillard and D.V.Nanopoulos, Unification of the fundamental particle interactions. Ed. A. Zichichi (Plenum) 1980, 461.
17) For a compendium, see
A.C.Scott, F.Y.F.Chu and D.W.McLaughlin, Proc. IEEE 61(1973) 1443.
18) For a review see, for example
D. Olive, Rivista del Nuovo Cimento 2 (1979) 1.
M.F.Atiyah, Geometry of Yang-Mills Fields, Lezione Fermioni Pisa (1979).
E.F.Corrigan and P. Goddard, Lecture Notes in Physics 129. Eds. J.P.Harnad and S. Schnider (Springer-Verlag) 1980.
19) E.B.Bogomolny, Sov. J. Nucl. Physics 24 (1976) 449.
20) M.K.Prasad and C.M.Sommerfield, Phys. Rev. Lett. 35 (1975) 76.
S. Coleman, S. Powke, A. Neveu and C.M.Sommerfield, Phys. Rev. D15 (1977) 544.
21) N.S.Manton, Nucl. Phys. B126 (1977) 525.
22) A. Jaffe and C. Taubes, Vortices and Monopoles (Birkhauser, Boston) 1980.
23) E. Weinberg, Phys. Rev. D20 (1979) 936.
24) P. Houston and L. O'Raifeartaigh, Phys. Lett. 93B (1980) 151, 94B (1980) 153.
25) P. Forgacs, Z. Horvath and L. Palla, Phys. Lett. B99 (1981) 232.
B102 (1981) 131.
26) N. Manton, Nucl. Phys. B135 (1978) 319.
For a review see,
L. O'Raifeartaigh and S. Rouhani, Recent Developments in Finite Energy (Topological) monopole theory, Schladming (1981).
27) F.J.Ernst, Phys. Rev. 167 (1968) 1175.
28) B.K.Harrison, Phys. Rev. Lett. 41 (1978) L67.
G. Neugebauer, J. Phys. A12 (1979) L67.
29) See for example,
D. Olive, Classical solutions in Gauge Theories - spherically symmetric monopoles - Lax Pairs and Toda Lattices. Lectures given at the International Summer Institute on Theoretical Physics Bad Honnef (1980) (to be published by Plenum).

or, more recently,
N. Ganoulis, P. Goddard and D. Olive, 'Self Dual Monopoles and Toda Molecules', Imperial College preprint (1981).
30) C.N.Yang, Phys. Rev. Lett. 39 (1977) 1377.
31) E.F.Corrigan, D.B.Fairlie, P. Goddard and R. Yates, Comm. Math. Phys. 58 (1978) 2528.
32) M.K.Prasad, Comm. Math. Phys. 80 (1981) 137.
33) For some more details see
E.F.Corrigan, Multimonopoles, Lectures given at the meeting on Integrable quantum field theories, Trarminne, Finland (1981).
34) R. Hirota, J. Phys. Soc. Japan 35 (1973) 289.
35) See, for example
Advances in Twistor Theory. Eds. L.P.Hughston and R.S.Ward (Pitman) 1979.
36) R.S.Ward, Phys. Lett. 61A (1977) 81.
37) M.F.Atiyah and R.S.Ward, Comm. Math. Phys. 55 (1977) 117.
38) R.S.Ward, Comm. Math. Phys. 79 (1981) 317.
39) E.F.Corrigan and P. Goddard, Comm. Math. Phys. 80 (1981) 575.
40) M.K.Prasad, A. Sinha and L.L.Chan Wang, Phys. Rev. D23 (1981) 2321.
M.K,Prasad and P. Rossi, Phys. Rev. Lett. 46 (1981) 806.
41) The separated two monopole was spotted by
R.S.Ward, Phys. Lett. B102 (1981) 136, Comm. Math. Phys. 80 (1981) 563.
42) W. Nahm, 'All self dual monopoles for arbitrary gauge groups' (CERN preprint 3172). Talk given at the International Summer Institute on Theoretical Physics Freiburg (1981).

YANG-MILLS THEORY AND GRAVITATION: A COMPARISON*

Andrzej Trautman**

Institut für Theoretische Physik
Universität Wien

Introduction

The purpose of theoretical physics is to construct mathematical models of physical phenomena and, on the basis of such models, to explain what is observed, suggest new experiments and predict their outcome. This ideal activity is supplemented and supported by research, done by mathematicians and physicists, on the properties of the theoretical constructions themselves. One considers questions such as: Are the equations of a theory consistent? Are their solutions stable? Can the Cauchy problem be correctly formulated? In what space of functions? Answers to such questions have no direct bearing on the predictive power of a theory, but they may throw light on the range of its applicability or need for modifications. Successful physical theories are often studied in order to construct, by analogy, models of phenomena outside their scope. For example, in the 19th century, there was a trend to reduce all of physics to classical mechanics, whereas now quantum electrodynamics is the theory relative to which all others are evaluated.

The quantum-mechanical description of charged particles led to an important change in the original interpretation, due to Weyl [1], of gauge transformations as maps inducing conformal changes of the metric tensor in space-time. The idea that the electromagnetic field is a 'compensating' or 'gauge' field [7] associated with the circle group U(1) was generalized, by Yang and Mills [10], by the introduction of a gauge field corresponding to the 'isotopic' group SU(2). Soon after, it became clear that essentially any Lie group can be so 'gauged' and that Einstein's theory of gravitation fits - though not quite - into the scheme (cf. the Annotated Bibliography for references and further remarks on the history of the development of the notion of a gauge field).

Present-day physics is dominated by the striking successes of quantum electrodynamics and the current trends in the description of fundamental interactions

* The actual lectures given by the author at the 1981 Scheveningen Conference contained, besides the material reproduced here, an introduction to the geometrical aspects of gauge theories, based on articles published elsewhere [52,61].

** Permanent address: Instytut Fizyki Teoretycznej, Uniwersytet Warszawski, Hoża 69, Warszawa, Poland.

(chromodynamics and the Weinberg-Salam theory). As a result of this, the theory of gravitation is sometimes required to conform to the principles and fashions prevalent in elementary particle physics. In my opinion, one should rather regard Einstein's theory of general relativity in its own right, as a very successful, classical, relativistic theory of gravitation. Its structure is worth studying and comparing to that of theories of the Yang-Mills type, but not necessarily with an intention of formulating all gauge theories according to one pattern. If a unified picture is desired, it may be achieved not so much by replacing Einstein's equations by those arising from a Lagrangian quadratic in the field strengths, as by combining gravitation with Yang-Mills fields in a suitably generalized theory of the Kaluza-Klein type [12,18,22-25,27,44,57].

Superficial Observations

Consider the following three classical, relativistic field theories:

(i) Maxwell's electrodynamics,

(ii) Yang-Mills theory based on SU(2),

(iii) Einstein's theory of gravitation.

They share some fundamental properties: on the mathematical side, each of the theories is based on an infinitesimal connection defined on a suitable principal bundle over space-time; they all exhibit 'large' groups of gauge transformations. From the point of view of physics, the similarities between (i) and (iii) are obvious: the Coulomb law is analogous to Newton's. In fact, electromagnetism and gravitation seem to be the only two long range forces existing in nature. Free Yang-Mills equations also have Coulomb-like solutions, but their physical relevance is probably restricted by the phenomenon of confinement and/or the Higgs-Kibble mechanism of mass generation through a spontaneous breakdown of symmetry.

A superficial analysis indicates analogies between (i) and (ii), as well as between (ii) and (iii), but not so much between (i) and (iii). Since the Maxwell and Yang-Mills Lagrangians are both quadratic in the field strengths, they yield equations of a similar form. On the other hand, Yang-Mills and Einstein equations exhibit non-linearities which, in both cases, may be traced back to the non-Abelian character of the corresponding structure groups. These non-linearities induce a self-interaction of the corresponding particles.

In the rest of the paper, the analogies and differences among the three theories (i) - (iii) will be considered and some unexpected formal similarities between gravitation and electromagnetism stressed.

A Dictionary

Much of the language of theoretical physics is sufficiently imprecise to allow vivid disputes between authors who attribute different meanings to the words they use. This is the way it has to be: the exact significance of the notions used in physics becomes clear only in the final stages of formation of the theories in which they occur. It is now being felt that classical gauge theories have reached the point when their fundamental notions can be given a precise meaning, i.e. translated into clearly defined mathematical terms. Such a dictionary has been initiated by Wu and Yang [26] and I supplement it here with a few entries.

A classical gauge theory is any physical theory which includes among its dynamical variables a connection on a principal G-bundle P over space-time M. The structure group G is a Lie group; physicists often call it the 'gauge group', but this is misleading as the same name is used (more appropriately) for a group of automorphisms of the bundle. In the physicist's language 'to gauge a group G' means 'to consider (sometimes: to construct) a connection on a bundle over space-time with structure group G'. A connection form ω on π: $P \to M$ describes a 'gauge configuration' and a local section s: $U \to P$, $U \subset M$, $\pi \circ s = \mathrm{id}$, defines a 'gauge'. The pull-back $A = s^*\omega$ is the 'potential of the gauge configuration in the gauge s'. Similarly, if $\Omega = d\omega + \frac{1}{2}[\omega,\omega]$ is the curvature two-form on P, then $F = s^*\Omega$ is the 'field strength in the gauge s'.

Let M be an oriented Riemannian space (conformal geometry suffices if M is four-dimensional) and let $*$ denote the Hodge (duality) isomorphism of the vector structure of the Grassmann algebra over M. This isomorphism lifts to horizontal forms on P. Let k: $\mathfrak{g} \times \mathfrak{g} \to \mathbb{R}$ be a scalar product on the Lie algebra $\mathfrak{g}$ of G, invariant under the adjoint action of G in $\mathfrak{g}$. If (e_i) is a linear basis in $\mathfrak{g}$, $k_{ij} = k(e_i,e_j)$ and $\Omega = \Omega^i e_i$, then

$$k_{ij} *\Omega^i \wedge \Omega^j \qquad (1)$$

is a G-invariant, horizontal form of degree n = dim M. The pull-back of (1) with a section s: $M \to P$ does not depend on s; upon integration over M it gives the classical action from which field equations are derived by variation. A gauge theory is said to be of the 'Yang-Mills type' if its action contains a term derived from (1). If G is semi-simple and compact, then k may be taken as a multiple of its Killing-Cartan form: this is the case of the 'Yang-Mills theory'. For example, Maxwell's electrodynamics is a theory of the Yang-Mills type, but not a Yang-Mills theory in the strict sense. Einstein's general relativity is not a theory of the Yang-Mills type (see, however, [28,38,45-47] for different views on this problem).

Some Analogies and Differences

It is convenient to summarize the analogies and differences between gauge theories of the Yang-Mills type and gravitation, some of which have already been mentioned, in a table where the following notation is used:

$\theta = (\theta^\mu)$ is the canonical (soldering) $\mathbb{R}^n$-valued 1-form on the bundle $LM \to M$ of linear frames of an n-dimensional manifold M, μ and other Greek indices run from 1 to n;

$\omega = (\omega^\mu{}_\nu)$ is the 1-form of a linear connection;

$\Gamma = (\Gamma^\mu{}_\nu)$ are its coefficients, obtained by pull-back of ω by a (local) section $s: M \to LM$;

$e^\mu = s^*\theta^\mu$ is the μ-th element of the coframe field on M, dual to the frame field $s = (s_\mu)$, i.e. $\langle s_\mu, e^\nu\rangle = \delta^\nu_\mu$;

D denotes the covariant exterior derivative; if ϕ is a V-valued field of k-forms of type ρ, defined by a homomorphism $\rho: \mathcal{g} \to L(V)$ of Lie algebras, then $D\phi =$ $= d\phi + \rho(A) \wedge \rho$ [49];

$R = (R^\mu{}_\nu)$ is the curvature two-form, referred to the frame s, $R = s^*\Omega$, where $\Omega^\mu{}_\nu = d\omega^\mu{}_\nu + \omega^\mu{}_\rho \wedge \omega^\rho{}_\nu$;

$Q = (Q^\mu)$ is the torsion two-form, referred to the frame s, $Q = s^*\Theta$, where $\Theta^\mu = d\theta^\mu + \omega^\mu{}_\nu \wedge \theta^\nu$;

g is the metric tensor and $g_{\mu\nu} = g(s_\mu, s_\nu)$;

$\eta_{\mu\nu}$ is the Hodge dual of $e_\mu \wedge e_\nu$, where $e_\mu = g_{\mu\nu}e^\nu$;

$\eta_{\mu\nu\rho}$ is the Hodge dual of $e_\mu \wedge e_\nu \wedge e_\rho$;

$T_\mu = T_{\mu\nu}e^\nu$ is the $\mathbb{R}^n$-valued 1-form of energy.momentum of the sources of the gravitational field; similarly, t_μ corresponds to the 'pseudotensor' of energy-momentum of the gravitational field itself;

j is the $\mathcal{g}$-valued 1-form of the current corresponding to the sources of the gauge field;

ϕ is a (generalized) Higgs field, i.e. a V-valued field of k-forms of type ρ.

The most important difference between theories of the Yang-Mills type and gravitation is that the underlying bundle of the latter - the bundle of linear frames - is 'concrete', has more structure than 'abstract' bundles occurring in other gauge theories. The additional structure is completely characterized by the soldering form which, upon differentiation, leads to torsion. In Einstein's theory torsion is assumed to vanish. This condition has no counterpart in theories of the Yang-Mills type.

The role played by the metric tensor in Einstein's theory is somewhat analogous to that of a Higgs field in a Yang-Mills theory. In both cases the additional structure 'breaks down the symmetry' by restricting the principal bundle to a subgroup H of its structure group G. If $\phi: P \to V$ is a V-valued map, equivariant under the action of G in P and in V defined by a representation $\rho: G \to GL(V)$, and such that the values

Table

Yang-Mills		Gravitation
A		Γ
F		R
$DF = 0$	Bianchi identity	$DR = 0$
—	torsion	Q
Higgs field ϕ		metric tensor g
$D\phi = 0$	compatibility	$Dg = 0$
$k_{ij} *F^i \wedge F^j$	field Lagrangian	$\eta^{\nu}{}_{\mu} \wedge R^{\mu}{}_{\nu}$
$D*F = 4\pi * j$	field equations	$\frac{1}{2}\eta_{\mu\nu\rho} \wedge R^{\nu\rho} = -8\pi * T_\mu$
$d*F = 4\pi * j - [A,*F]$	field equations in Gauss's form	$dU_\mu = 4\pi*(T_\mu + t_\mu)$, where $U_\mu = \frac{1}{4}\eta_{\mu\nu\rho} \wedge \Gamma^{\nu\rho}$
$d(*j - \frac{1}{4\pi}[A,*F]) = 0$	conservation law	$d*(T_\mu + t_\mu) = 0$
$d*A = 0$	gauge fixing condition	$d*e = 0$

Boundary conditions at spatial infinity for static configurations

$\phi = O(1)$		g = Minkowski tensor + $O(1/r)$
$A = O(1/r)$		$\Gamma = O(1/r^2)$
$F = O(1/r^2)$		$R = O(1/r^3)$
$\frac{1}{4\pi}\oint *F$	total conserved quantity	$\frac{1}{4\pi}\oint U_\mu$

of ϕ lie in an orbit $W \subset V$ of G, then H is the isotropy group of some point of $W \simeq G/H$. In general, there are many orbits in V corresponding to the same H: they are all said to belong to the same stratum. For example, in a standard SO(3) Yang-Mills-Higgs theory, under the assumption of spherical symmetry and $\phi \neq 0$, the normalized field $\phi/||\phi||$ breaks the symmetry down to $H = SO(2)$. The radial Higgs equation selects, for each radius r, an orbit containing $\phi(r) \in \mathbb{R}^3$. All these orbits are diffeomorphic to $S_2 \simeq SO(3)/SO(2)$: they belong to the same stratum, without being isometric [48]. The situation is rather different in the theory of gravitation, where $G = GL(n,R)$ and H is an orthogonal group. According to the 'theorem on inertia' of quadratic forms, each stratum in $L^2_s(\mathbb{R}^n,\mathbb{R})$ consists of a single orbit, viz. the set of all quadratic forms with a given signature. As a result of this, there is no 'radial equation' and potential for the metric tensor; the symmetry breaking in the theory of gravitation is more of kinematic than dynamic nature.

An essential difference between the two types of theories occurs in connection with the asymptotic behaviour (at large distances) of their static fields; this is indicated in the Table. A gauge transformation of the potential, $A \to A'$,

$$A' = S^{-1} A S + S^{-1} dS ,$$

where

$$S: M \to G ,$$

is compatible with the asymptotic behaviour of a time-independent A, if

$$S = a(\theta,\phi)(I + \beta(\theta,\phi)/r + \ldots) ,$$

where θ, ϕ are coordinates on S_2, and $a: S_2 \to G$. Under such a transformation, the field strenghts change as follows,

$$F' = a^{-1} F a + O(1/r^3) .$$

Therefore, the total non-Abelian charge

$$\frac{1}{4\pi} \oint *F$$

is ill-defined [34]. By contrast, in the theory of gravitation, one has $\Gamma = O(1/r^2)$ for static configurations. To preserve this asymptotic behaviour, in the generic case, one has to restrict $a = (a^\mu{}_\nu)$ to be a constant matrix. This allows one to define unambiguously the total mass for such configurations. Indeed, the Von Freud 'superpotential' U transforms as follows,

$$U'_\mu = U_\nu a^\nu{}_\mu + O(1/r^3) , \qquad \text{where } a \in SO(1,3) .$$

The structure of the group of gauge transformations also reflects the similarities and differences among gauge theories [55]. A gauge transformation is an automorphism of the principal bundle $\pi: P \to M$ preserving the absolute elements of the gauge theory. A gauge transformation is said to be pure if it is vertical (based), i.e. if it induces the identity map on M. For any gauge theory one can construct the (horizontally) exact sequences of group homomorphisms,

$$\begin{array}{ccccccccc} I & \to & \mathcal{G}_0 & \to & \mathcal{G} & \to & \mathcal{G}/\mathcal{G}_0 & \to & I \\ & & \downarrow & & \downarrow & & \downarrow & & \\ I & \to & \mathrm{Aut}_0\, P & \to & \mathrm{Aut}\, P & \to & \mathrm{Diff}\, M & & \end{array}$$

where $\mathcal{G}$ (resp. $\mathcal{G}_0$) is the group of all gauge (resp. all pure gauge) transformations and Aut P (resp. Aut_0 P) is the group of all (resp. all vertical) automorphisms of P. In general relativistic theories of gravitation, the soldering form on P = LM is an absolute element and it reduces $\mathcal{G}$ to Diff M and $\mathcal{G}_0$ to the identity. By contrast, in a theory of the Yang-Mills type over Minkowski space, both $\mathcal{G}_0$ and $\mathcal{G}$ are 'large' groups, but $\mathcal{G}/\mathcal{G}_0$ is 'small', i.e. a Lie group [61].

Plane Gravitational Waves Are Abelian

Another aspect of Einstein's theory of gravitation, which makes it resemble electrodynamics rather than non-Abelian Yang-Mills theories, is associated with the nature of its plane waves.

In any theory of the Yang-Mills type, the potential

$$A = (a(u)x + b(u)y + c(u))du \,, \qquad (2)$$

where $u = t - z$ and $a,b,c\colon \mathbb{R} \to \mathfrak{g}$, represents in Minkowski space a solution of the source-free equation $D{*}F = 0$. The corresponding field strength

$$F = (adx + bdy) \wedge du$$

is invariant under translations in the (x,y)-plane, but the potential - and therefore the entire gauge configuration - is not, in general. For example, for $G = SO(3)$ and $[a,b] \neq 0$ the potential (2) is not invariant under any translation in that plane. On the other hand, if the functions a, b, and c span an Abelian Lie subalgebra of $\mathfrak{g}$, then (2) is invariant under translations in the (x,y)-plane and c can be eliminated by a gauge transformation.

The connection form Γ of plane gravitational waves, referred to a suitable orthonormal frame, can also be written in the form (2). In this case, however, the functions $a,b,c\colon R \to so(1,3)$ span a two-dimensional, Abelian subalgebra n of $so(1,3)$, corresponding to the nilipotent part of its Iwasawa decomposition. Therefore, c can be eliminated and the solution has a 5-dimensional group of isometrics isomorphic to the group of symmetries of a plane electromagnetic wave propagating in one direction. Incidentally, the restriction to n of the polarizational degrees of freedom is a result of the vanishing of torsion. There does not seem to exist an analogous, natural restriction on a and b in the non-Abelian Yang-Mills theory.

Acknowledgments

This text has been written in November, 1981, during a visit to the Institut für Theoretische Physik der Universität Wien. I thank P. Aichelburg, R. Beig, H. Grosse, R. Sexl, W. Thirring and H. Urbantke for their hospitality and discussions. A grant from the Einstein Memorial Foundation, which made possible my stay in Vienna, is gratefully acknowledged.

Annotated Bibliography

[1] H. Weyl, Gravitation und Elektrizität, Sitzungsber. Preuss. Akad. Wiss. (1918) 465.
The notion of 'gauge transformations' is introduced here, for the first time, in connection with an attempt to unify gravitation and electromagnetism. Gauge transformations act on the metric of space-time.

[2] Th. Kaluza, Zum Unitätsproblem der Physik, Sitzungsber. Preuss. Akad. Wiss. (1921) 966.
Kaluza shows that the Einstein-Maxwell equations can be geometrically interpreted in a five-dimensional Riemannian space whose metric is independent on one of the coordinates, say x^5. Gauge transformations are reduced to coordinate changes, $\bar{x}^i = x^i$ $(i = 1,2,3,4)$ and $\bar{x}^5 = x^5 + f(x^1,x^2,x^3,x^4)$.

[3] E. Schrödinger, Über eine bemerkenswerte Eigenschaft der Quantenbahnen eines einzelnen Elektrons, Zeitschrift f. Phys. 12 (1922) 13-23.

[4] E. Cartan, Sur les variétés à connexion affine et la théorie de la relativité généralisée, Ann. Ecole Norm. 40 (1923) 325-412; 41 (1924) 1-25 and 42 (1925) 17-88.
General relativity is extended and slightly modified by admitting a metric linear connection whose torsion tensor is related to the density of intrinsic angular momentum.

[5] O. Klein, Quantentheorie und fünfdimensionale Relativitätstheorie, Zeitschrift f. Phys. 37 (1926) 895.
Klein extends the theory of Kaluza [2] by allowing a periodic dependence of the field variables on the fifth coordinate.

[6] F. London, Quantenmechanische Deutung der Theorie von Weyl, Zeitschrift f. Phys. 42 (1927) 375.

[7] H. Weyl, Elektron und Gravitation I, Zeitschrift f. Phys. 56 (1929) 330-352.
Under the influence of the development of quantum mechanics [3,6], Weyl abandons his earlier interpretation of gauge transformations and accepts the one that connects a change in the electromagnetic potential to a transformation of the wave function of a charged particle.

[8] H. Weyl, A remark on the coupling of gravitation and electron, Phys. Rev. 77 (1950) 699-701.
By varying the Einstein-Dirac action integral independently with respect to the metric tensor and the components of a (metric) linear connection, one arrives at a set of equations closely related to those considered by Cartan [4].

[9] Ch. Ehresmann, Les connexions infinitésimales dans un espace fibré différentiable, in: Coll. de Topologie (Espaces Fibrés), Bruxelles, 5-8 juin 1950, G. Thone, Liège et Masson, Paris, 1951.
An (infinitesimal) connection is defined as an invariant distribution of horizontal linear spaces on the total space of a differentiable principal bundle. Cartan connections are defined in terms of soldering.

[10] C.N. Yang and R.L. Mills, Conservation of isotopic spin and isotopic gauge invariance, Phys. Rev. 96 (1954) 191.
The fundamental paper where gauge transformations are extended to the SU(2) group and quantization of the non-Abelian gauge field is considered.

[11] T.D. Lee and C.N. Yang, Conservation of heavy particles and generalized gauge transformations, Phys. Rev. 98 (1955) 1501.
The authors conjecture that all internal conserved quantities arise from invariance under gauge transformations.

[12] O. Klein, Generalisations of Einstein's theory of gravitation considered from the point of view of quantum theory, Helv. Phys. Acta, Suppl. 10 (1956) 58.

[13] R. Utiyama, Invariant-theoretical interpretation of interaction, Phys. Rev. 101 (1956) 1597.
Gauge transformations and potentials are defined for an arbitrary Lie group. Gravitation is interpreted as a gauge theory of the Lorentz group.

[14] M. Gell-Mann, The interpretation of the new particles as displaced charge multiplets, Nuovo Cimento 4, Suppl. 2 (1956) 848.
The principle of minimal coupling is formulated for electromagnetic interactions.

[15] D.W. Sciama, On the analogy between charge and spin in general relativity, in: Recent Developments in General Relativity (Infeld's Festschrift) pp. 415-439,

Pergamon Press and PWN, Oxford and Warsaw, 1962.
Weyl's ideas [7,8] are extended to obtain a coupling between classical spin and a linear connection with torsion. By analogy with electromagnetism [14], a principle of minimal coupling is formulated for the gravitational field.

[16] T.W.B. Kibble, Lorentz invariance and the gravitational field, J. Math. Phys. 2 (1961) 212.
The gravitational field is treated as a gauge field. Full use is made of both 'translations' (diffeomorphisms of space-time) and Lorentz transformations (rotations of the orthonormal frames in tangent spaces). The field equations, derived from a variational principle of the Palatini type, lead to a theory with a connection which is metric, but not necessarily symmetric. References are given to Weyl [8], Yang and Mills [10], Utiyama [13] and Sciama [15].

[17] E. Lubkin, Geometric definition of gauge invariance, Annals of Phys. 23 (1963) 233-283.
The significance of fibre bundles with connections to describe gauge configurations is recognized and used to relate the quantization of dual (magnetic) charges to the homotopy classification of bundles. Analogies between gravitation and Yang-Mills fields are stressed.

[18] B.S. DeWitt, article in Relativité, groupes et topologie, p. 725, edited by C. DeWitt and B.S. DeWitt, Gordon and Breach, New York, 1964.
The Kaluza-Klein construction is generalized to non-Abelian gauge fields.

[19] F. Hehl and E. Kröner, Über den Spin in der allgemeinen Relativitätstheorie: Eine notwendige Erweiterung der Einsteinschen Feldgleichungen, Zeitschrift f. Phys. 187 (1965) 478-489.
General relativity with spin and torsion is shown to resemble a field theory of dislocations.

[20] B. O'Neill, The fundamental equations of a submersion, Michigan Math. J. 13 (1966) 459-469.
A natural Riemannian metric is defined on the total space of a principal bundle with connection over a Riemannian base. The curvature of such a 'generalized Kaluza-Klein' metric is computed. This work has been extended by Gray [21] and Jensen [25].

[21] A. Gray, Pseudo-Riemannian almost product manifolds and submersions, J. Math. Mech. 16 (1967) 715-737.

[22] A. Trautman, Fibre bundles associated with space-time, Lectures at King's College, London, September 1967; published in Rep. Math. Phys. (Toruń) 1 (1970) 29-62.
Description of a natural isomorphism between the generalized Kaluza-Klein space [18] and the total space of a principal bundle underlying a Yang-Mills theory. The notions of naturality and relativity are related one to another.

[23] R. Kerner, Generalization of the Kaluza-Klein theory for an arbitrary non-Abelian gauge group, Ann. Inst. Henri Poincaré 9 (1968) 143.
The system of Einstein-Yang-Mills equations is derived from a geometric principle of least action. The author misses a 'cosmological' term arising in the non-Abelian case.

[24] W. Thirring, Remarks on five-dimensional relativity, in: The Physicist's Conception of Nature, pp. 199-201, edited by J. Mehra, D. Reidel, Dordrecht, 1973.
The Dirac equation on the Kaluza-Klein space leads, in a natural way, to CP-violation.

[25] G.R. Jensen, Einstein metrics on principal fibre bundles, J. Diff. Geometry 8 (1973) 599-614.

[26] T.T. Wu and C.N. Yang, Concept of nonintegrable phase factors and global formulation of gauge fields. Phys. Rev. D12 (1975) 3845-3857.
Connections on principal bundles are recognized to be relevant for the descriptiom of topologically non-trivial gauge configurations, such as those arising in the Bohm-Aharonov experiments. Dirac's quantization condition for the magnetic monopole is shown to be equivalent to the classification of U(1)-bundles over S_2 in terms of their first Chern classes.

[27] Y.M. Cho, Higher-dimensional unifications of gravitation and gauge theories, J. Math, Phys. 16 (1975) 2029-2035.
The generalized Kaluza-Klein theory is developed and Kerner's mistakes [23] are corrected.

[28] Y.M. Cho, Einstein Lagrangian as the translational Yang-Mills Lagrangian, Phys. Rev. D14 (1976) 2521-2525.
[29] H.J. Bernstein and A.V. Phillips, Fiber bundles and quantum theory, Scientific American 245 (1981) 123-137.
The ultimate sign of acceptance: Scientific American publishes an article on the geometry of gauge fields.

Some Recent Reviews

[30] E.S. Abers and B.W. Lee, Gauge theories, Phys. Rep. 9 (1973) 1-41.
[31] S. Weinberg, Recent progress in gauge theories of the weak, electromagnetic and strong interactions, Rev. Mod. Phys. 46 (1974) 255-277.
[32] L.D. Faddeev, Differential-geometric structures and quantum field theory, Trudy Mat. Inst. Steklova 135 (1975) = Proc. Steklov Inst. Math. 1 (1978) 223-228.
[33] C.H. Gu and C.N. Yang, Some problems on the gauge field theories, Scientia Sinica, Part I: 18 (1975) 483-501; Part II: 20 (1977) 47-55; Part III: 20 (1977) 177-185.
[34] B.D. Bramson, Relativistic angular momentum for asymptotically flat Einstein-Maxwell manifolds, Proc. Roy. Soc. (Lond.) A341 (1975) 463-490.
[35] J.C. Taylor, Gauge theories of weak interactions, Cambridge University Press, Cambridge, 1976.
[36] F.W. Hehl, P. Von der Heyde, G.D. Kerlick, and J.M. Nester, General relativity with spin and torsion: Foundations and prospects, Rev. Mod. Phys. 48 (1976) 393-416.
[37] A. Trautman, A classification of space-time structures, Rep. Math. Phys. (Toruń) 10 (1976) 297-310.
[38] F. Mansouri and L.N. Chang, Gravitation as a gauge theory, Phys. Rev. D13 (1976) 3192-3200.
[39] W. Drechsler and M.E. Mayer, Fiber bundle techniques in gauge theories, Lecture Notes in Phys. No 67, Springer, Berlin, 1977.
[40] J.P. Harnad and R.B. Pettitt, Gauge theory of the conformal group, in: Group-theor. methods in physics, Proc. of the Fifth Intern. Colloquium, Academic Press, New York, 1977.
[41] R. Stora, Continuum gauge theories, in: New Developments in Quantum Field Theory and Statistical Mechanics, edited by M. Lévy and P. Mitter, Plenum, New York, 1977.
[42] M.F. Atiyah, Geometrical aspects of gauge theories, Proc. Intern. Congress Math., vol II, pp. 881-885, Helsinki, 1978.
[43] A. Jaffe, Introduction to gauge theories, ibid., pp. 905-916.
[44] R. Hermann, Yang-Mills, Kaluza-Klein and the Einstein program, Math. Sci. Press, Brookline, Mass., 1978.
[45] Y. Ne'eman and T. Regge, Gauge theory of gravity and supergravity on a group manifold, Riv. Nuovo Cimento 1: 5 (1978) 1.
[46] W. Thirring, Gauge theories of gravitation, Lecture at XVII Universitätswochen für Kernphysik (Schladming, 1978), Acta Phys. Austr., Suppl. XIX (1978) 439.
[47] P. Van Nieuwenhuizen and D.Z. Freedman (eds), Supergravity (Proc. of the Supergravity Workshop at Stony Brook, 1979) North-Holland, Amsterdam, 1979.
[48] L. O'Raifeartaigh, Hidden gauge symmetry, Rep. Prog. Phys. 42 (1979) 159-224.
[49] A. Trautman, The geometry of gauge fields, Czech. J. Phys. B29 (1979) 107-116.
[50] M. Daniel and C.M. Viallet, The geometrical setting of gauge theories of the Yang-Mills type, Rev. Mod. Phys. 52 (1980) 175-197.
[51] T. Eguchi, P.B. Gilkey, and A.J. Hanson, Gravitation, gauge theories and differential geometry, Phys. Rep. 66 (1980) 213-393.
[52] A. Trautman, Fibre bundles, gauge fields, and gravitation, in: General Relativity and Gravitation, vol. I, pp. 287-308, edited by A. Held, Plenum, New York, 1980.
[53] F.W. Hehl, J. Nitsch, and P. Von der Heyde, Gravitation and the Poincaré gauge field theory with quadratic Lagrangian, ibid., pp. 329-355.
[54] R. Jackiw, Introduction to the Yang-Mills quantum theory, Rev. Mod. Phys. 52 (1980) 661-673.
[55] A. Trautman, On groups of gauge transformations, Lecture Notes on Phys. No 129, pp. 114-120, Springer, Berlin, 1980.

[56] J. Iliopoulos, Unified theories of elementary particle interactions, Contemp. Phys. 21 (1980) 159-183.

[57] W. Kopczyński, A fibre bundle description of coupled gravitational and gauge fields, Lecture Notes in Math. No 836, p. 462, Springer, Berlin, 1980.

[58] H. Woolf (editor), Some Strangeness in the Proportion: A Centennial Symposium to Celebrate the Achievements of Albert Einstein, Addison-Wesley, Reading, Mass., 1980.

[59] G.H. Thomas, Introductory lectures on fibre bundles and topology for physicists, Riv. Nuovo Cimento 3: 4 (1980) 1-119.

[60] G. Mack, Physical principles, geometrical aspects, and locality properties of gauge field theories, Fortschritte d. Physik, 29 (1981) 135-185.

[61] A. Trautman, Geometrical aspects of gauge configurations, Lectures at XX Universitätswochen für Kernphysik (Schladming, 1981), Acta Phys. Austr., Suppl. XXIII (1981) 401-432.

THE TWISTOR DESCRIPTION OF LINEAR FIELDS

Michael G. Eastwood
Mathematical Institute
Oxford University, OXI 3LB,
England

After introducing twistors [19], Penrose used them to give integral formulae for massless free fields on Minkowski space [20]. After much refinement by Penrose [22, 24], these formulae were seen by Ward [28] to generalize to certain non-linear equations. These generalizations have proved very successful [1,2,3,29] but even for linear equations the approach is useful. The purpose of these lectures is to give a revue of the present status of the linear theory. Although I start from scratch the presentation is rather dense. Further details and motivation regarding the introductory material can be found in [14,23,30].

Spinors : Suppose S is a 2-dimensional complex vector space and denote by $\overline{S}$ its complex conjugate i.e. as Abelian groups $\overline{S} = S$ but the action of $\mathbb{C}$ is replaced by the conjugate action : if $s \in S$ and we write $\overline{s}$ for the corresponding element of $\overline{S}$ then $\lambda\overline{s} = \overline{\overline{\lambda}s}$ for $\lambda \in \mathbb{C}$. Form $V = S \otimes_{\mathbb{C}} \overline{S}$ and let $\overline{} : V \to V$ be defined by $s \otimes \overline{t} \mapsto t \otimes \overline{s}$. This is an $\mathbb{R}$-linear automorphism with $(\overline{})^2 = 1$ and so the "Hermitian" elements of V (those preserved by $\overline{}$) form a 4-dimensional $\mathbb{R}$-subspace L . Now suppose $\langle\ ,\ \rangle$ is a skew form on S . Then we obtain a symmetric form $(\ ,\ )$ on V by $(r \otimes \overline{s}\ ,\ t \otimes \overline{u}) = \langle r,t\rangle\ \langle s,u\rangle$. This form is $\mathbb{R}$-valued on L and it is easy to check that its signature is $+---$. Note that $\langle\ ,\ \rangle$ can only be changed by scale say $\langle\ ,\ \rangle \mapsto \Omega\langle\ ,\ \rangle$. Then we see that $(\ ,\ ) \mapsto |\Omega|^2(\ ,\ )$. Thus, even if we do not specify $\langle\ ,\ \rangle$ we still obtain a conformal metric on L with Lorentzian signature. If $A : S \to S$ is $\mathbb{C}$-linear preserving $\langle\ ,\ \rangle$ then $A \otimes \overline{A} : V \to V$ preserves $(\ ,\ )$ and so we obtain a homomorphism $SL(2,\mathbb{C}) \to O_+^\uparrow(1,3)$ = the connected component of the identity of the Lorentz group. This homomorphism is a double covering. Suppose we now identity L with the tangent space at some point of a Lorentzian manifold. Then the future-pointing null vectors are exactly $s \otimes \overline{s}$ for $s \in S$. Changing s by scale $s \mapsto \lambda s$ alters $s \otimes \overline{s}$ by the positive scale $|\lambda|^2$ and we see that this identifies the celestial sphere with the Riemann sphere $\mathbb{P}(S)$ and the Lorentz group $O_+^\uparrow(1,3)$ with the group $PSL(2,\mathbb{C})$ of holomorphic transformations of the celestial sphere $\mathbb{P}(S)$. This "rigidity" in complex analysis i.e. the fact that there are rather few holomorphic transformations of the sphere, is part of a recurrent theme in twistor theory. This observation is the first indication that complex analysis might be useful in general relativity.

Twistors : In general we may locally identify the tangent bundle of a Lorentzian manifold with the trivial bundle with fibre L but there is a topological obstruction to global consistency (a Stiefel-Whitney class). Such a global choice, i.e. a lift of the tangent bundle as an $O_+^{\uparrow}(1,3)$-bundle to an $SL(2,\mathbb{C})$-bundle, is called a spin structure. We will only be concerned with Minkowski space M^I for which this is rather trivial i.e. we fix an origin and identify $M^I = L$. Let us temporarily forget about the real structure and define complexified Minkowski space $\mathbb{M}^I$ as V $(=L \otimes \mathbb{C})$ with holomorphic metric $(\ ,\ )$. We may regard V as a space of linear transformations $V = \mathrm{Hom}(\overline{S}^*, S)$ or $\mathrm{Hom}(S^*, \overline{S})$. We choose one of these, say $V = \mathrm{Hom}(\overline{S}^*, S)$, and define twistor space as $T = S \oplus \overline{S}^*$. Note that there is a choice here and we will see that this leads to some asymmetry in the twistor description of fields but this is not necessarily a disadvantage since nature itself is not left-right symmetric (for example, neutrinos are always left-handed). If $x \in \mathbb{M}^I$ is regarded as a linear transformation $x : \overline{S}^* \to S$ then its graph $\{(\omega,\pi) \in T \text{ s.t. } \omega = x\pi\}$ is a 2-dimensional linear subspace of T. Hence $\mathbb{M}^I \subset Gr_2(T)$ = the Grassmannian of 2-dimensional linear subspaces of T. The point of doing this is that the conformal structure of $\mathbb{M}^I$ (i.e. the metric $(\ ,\ )$ but only up to scale) takes on a geometric significance in that two points $x,y \in \mathbb{M}^I$ are null separated if and only if the corresponding planes in T intersect in a line (rather than just the origin). To be slightly more economical let us form the projective space $\mathbb{P}T$, denoted from now on just by $\mathbb{P}$. Then points x of $\mathbb{M}^I$ give rise to lines L_x (linearly embedded Riemann spheres) in $\mathbb{P}$ and the conformal structure of $\mathbb{M}^I$ is characterized by whether these L_x and L_y intersect. Note that most (an open dense subset) lines in $\mathbb{P}$ arise in this way and so it is natural to compactify and define compactified complexified Minkowski space $\mathbb{M}$ as the space of all lines in $\mathbb{P}$ (i.e. $\mathbb{M} = Gr_2(T)$). The conformal structure extends to $\mathbb{M}$ and is characterized by null separation : two lines K and L are said to be null separated as points of $\mathbb{M}$ if and only if they intersect non-trivially in $\mathbb{P}$. At this stage we see how real Minkowski space may be recovered : recall $T = S \oplus \overline{S}^*$ and so carries a natural Hermitian form $\langle(\omega,\pi), (\mu,\theta)\rangle = i\overline{\theta}(\omega) - i\pi(\overline{\mu})$. If $x \in V$ happens to be real, i.e. $\overline{x} = x$, and $Z = (\omega,\pi) \in L_x$ (using homogeneous coordinates) then $\|Z\|^2 = \langle(\omega,\pi),(\omega,\pi)\rangle = i\overline{\pi}(x\pi) - i\pi(\overline{x\pi}) = 0$. Thus, if we denote by P the 5-real-dimensional submanifold of $\mathbb{P}$ on which $\|\ \|^2$ vanishes then the lines L_x corresponding to x real lie inside P.

Exercise : Let M denote the conformal compactification of real Minkowski space M^I i.e. the closure of M^I in $\mathbb{M}$. Describe M topologically. Show that the twistor correspondence $x \leftrightarrow L_x$ identifies M as the lines in $\mathbb{P}$ which lie inside P. Show that P may be regarded as the space of light rays (null geodesics) in M so that the twistor correspondence identifies a point in M with its celestial sphere. For the most part this additional real structure on $\mathbb{M}$ is ignored in what follows. There are other circumstances, for example the study of instantons on S^4, the conformal compactification of $\mathbb{R}^4$, which require a different real structure [1,2,3] .

Massless Fields : For the purposes of this lecture we can just take as definition $\mathbb{P} = Gr_1(T)$, $\mathbb{M} = Gr_2(T)$ for some 4-dimensional complex vector space T . To fix notation, and to define the massless field equations, we need to discuss the natural vector bundles which occur on $\mathbb{P}$ and $\mathbb{M}$.

Bundles on $\mathbb{P}$: There is the universal bundle which assigns to every point in $\mathbb{P}$ the line in T which that point represents. We shall denote this bundle by $\mathcal{O}(-1)$ and define $\mathcal{O}(n)$ more generally as $\mathcal{O}(-1)^{-n}$ for $n \in \mathbb{Z}$. If U is an open subset of $\mathbb{P}$ and $\widetilde{U}$ denotes the corresponding open subset of $T - \{0\}$ then there is a natural 1-1 correspondence $\Gamma(U,\mathcal{O}(n)) = \{$holomorphic functions $f(Z)$ on $\widetilde{U}$ s.t. $f(\lambda Z) = \lambda^n f(Z)$ for $\lambda \in \mathbb{C} - \{0\}\}$. For example, f s.t. $f(\lambda Z) = \lambda^{-1} f(Z)$ gives $f(Z)Z$ as a section of $\mathcal{O}(-1)$. This shows $\mathcal{O}(-1)$ is non-trivial since it has no non-trivial global sections.

Bundles on $\mathbb{M}$: Again there is the universal bundle. We shall denote it by $\mathcal{O}_{A'}$ and its dual by $\mathcal{O}^{A'}$. Letting $\mathcal{O}^{\alpha}$ denote the trivial bundle with fibre T we define $\mathcal{O}^A$ by means of the exact sequence $0 \to \mathcal{O}_{A'} \to \mathcal{O}^{\alpha} \to \mathcal{O}^A \to 0$ and $\mathcal{O}_A$ as its dual. This notation is explained more fully in [5] . The holomorphic tangent bundle to $\mathbb{M}$ is canonically identified with $\mathcal{O}^A \otimes \mathcal{O}^{A'}$ which we shall write from now on as $\mathcal{O}^{AA'}$. This identification is described in [18]. This global decomposition of the tangent bundle as a tensor product of two rank 2 vector bundles constitutes a spin structure and hence a conformal structure on $\mathbb{M}$. Since we are now dealing with complex objects there is no longer a complex conjugation relating $\mathcal{O}^A$ and $\mathcal{O}^{A'}$. For vector spaces, $A = B \otimes C \Rightarrow \wedge^2 A = (S^2 B \otimes \wedge^2 C) \oplus (S^2 C \otimes \wedge^2 B)$. Hence we can write 2-forms on $\mathbb{M}$ in terms of spinors : $\Omega^2 = \mathcal{O}_{(A'B')[AB]} \oplus \mathcal{O}_{(AB)[A'B']}$, where $\mathcal{O}_{(A'B')}$ denotes the symmetric tensor product $S^2\mathcal{O}_{A'}$ and $\mathcal{O}_{[A'B']}$ the skew tensor product $\wedge^2\mathcal{O}_{A'}$. This extends the splitting on M : $\Omega^2 = \Omega^2_+ \oplus \Omega^2_-$ where $\Omega^2_{\pm} = \{F \in \Omega^2$ s.t. $*F = \pm iF\}$; note that the Hodge star operator is conformally invariant (only depends on the metric up to scale) when acting on 2-forms. The source free Maxwell equations are $dF = 0$, $d*F = 0$ for a 2-form F . Splitting F into "self-dual" and "anti-self-dual" parts $F = F_+ + F_-$ the equations become $dF_+ = 0$, $dF_- = 0$. There are "non-Abelian" generalizations of Maxwell's equations (Yang-Mills equations) for which it is not possible to split a general solution like this but even then it is useful to look at the self-dual equations. In terms of spinors the self-dual Maxwell equations become $\nabla_{A[A'}F_{B']C'BC} = 0$ for $F_{B'C'BC} \in \Gamma(\mathcal{O}_{(B'C')[BC]})$. More generally, the massless free field equations for a helicity $n/2$ field $\phi_{B'C'\ldots D'BC} \in \Gamma(\mathcal{O}_{(B'\ldots D')[BC]})$ are $\nabla_{A[A'}\phi_{B']\ldots D'BC} = 0$ where ϕ has n primed indices. For negative n there are similar equations with primed and unprimed indices interchanged so that $dF_- = 0$ becomes the equations of helicity -1 and for $n = 0$ there is the wave equation. The following integral formulae for massless fields were discovered by Penrose [20]. They generalize the formula given in 1904 by Bateman for the wave equation. As described by Ward in these conference proceedings they can be modified to deal with the non-Abelian case which is a non-linear set of equations !

Integral formulae for massless fields : $\mathcal{O}_{[BC]}$ and $\mathcal{O}_{[B'C']}$ are isomorphic on $\mathbb{M}$ and we can choose an isomorphism so that the skew unprimed indices in the massless field equations are turned in primed indices. The equations for helicity 1/2 now read : $\nabla_{A[A'}\phi_{B'][C'D']} = 0$. Suppose f is a holomorphic function defined on some open subset of $\mathbb{T} - \{0\}$ and homogeneous of degree -3 i.e. $f(\lambda Z) = \lambda^{-3} f(Z)$. Then, simply by differentiating under the integral sign,
$\phi_{B'[C'D']}(x) = \oint \pi_{B'} f(x\pi,\pi)\pi_{[C'}d\pi_{D']}$ is a solution of the field equations. In this equation we have trivialized the bundle $\mathcal{O}_{B'}$ by means of the parametrization of the line L_x by $\pi \to (x\pi,\pi)$. This explains the meaning of the primed index on π . The formula is really independent of any choice of coordinates. What has not yet been explained is exactly where f is defined and over what set the integration takes place. For each fixed x the idea is to integrate over some contour located on the corresponding Riemann sphere L_x . The exact location of the contour is not important by Cauchy's theorem and explicit calculation is possible by residues. These considerations lead to :

The Penrose transform : An intuitive discussion of the relationship between the integral formulae and the cohomological reformulation is given in [24]. The idea is that f itself is not so important but that it represents something rather more natural, namely a cohomology class. It is clear that, whatever this means, the integral formula says that we should restrict it to the line L_x to find out what the value of the field at x is. A similar procedure can be carried out for functions : Suppose U is an open subset of M and U' is the corresponding open subset P i.e. the region swept out by the lines corresponding to points in U . Suppose $f \in \Gamma(U',\mathcal{O}(k))$. Then for each $x \in U$ we can restrict f to L_x to obtain an element of $\Gamma(L_x,\mathcal{O}(k))$. If $k = 0$ then, since L_x is compact, $f|_{L_x}$ must be constant. In this way we obtain a function on U but, since nearby lines in U' can be intersected by a third line, this function is necessarily constant. The case $k \geq 1$ is more interesting : here $\Gamma(L_x,\mathcal{O}(k))$ is a vector space of dimension k+1 which may be identified canonically with the symmetric k^{th} power of the dual of the plane in T which x represents. In other words we obtain a section of $\mathcal{O}^{(A'B'...D')}$. The geometry of intersecting lines constrains this section to be the restiction of one of the global sections $\Gamma(\mathbb{P},\mathcal{O}(k))$, a finite dimensional vector space of dimension $\binom{3+k}{k}$. This constraint can also be expressed locally as a differential equation
$\nabla_A^{(A'}\phi^{B'...D')} = 0$, the dual k-twistor equation. Like the massless field equations, these equations are also conformally invariant. Indeed, there are no other first order linear ones [8] . To motivate the introduction of sheaf cohomology let us consider further the vector space $\Gamma(L_x,\mathcal{O}(k))$. A good way of studying $\Gamma(L_x,\mathcal{O}(k))$ is

to pick two points "0" and "∞" on the sphere L_x, trivialize $\mathcal{O}(k)$ on $L_x - \{0\}$ and $L_x - \{\infty\}$, and expand elements of $\Gamma(L_x - \{0\}, \mathcal{O}(k))$ and $\Gamma(L_x - \{\infty\}, \mathcal{O}(k))$ as Taylor series (using $L_x - \{\infty\} \simeq \mathbb{C}$ etc.). A global section $f \in \Gamma(L_x, \mathcal{O}(k))$ can be characterized by comparing coefficients in these expansions to find $g \in \Gamma(L_x - \{\infty\}, \mathcal{O}(k))$ and $h \in \Gamma(L_x - \{0\}, \mathcal{O}(k))$ such that $g-h = 0$ on $(L_x - \{\infty\}) \cap (L_x - \{0\})$. In this way global sections are identified as polynomials of degree k. However there is a different way of comparing coefficients which looks at the "gap" between $\Gamma(L_x - \{0\}, \mathcal{O}(k))$ and $\Gamma(L_x - \{\infty\}, \mathcal{O}(k))$ rather than the "overlap". More precisely, $H^1(L_x, \mathcal{O}(k)) = \Gamma((L_x - \{\infty\}) \cap (L_x - \{0\}), \mathcal{O}(k))/C$ where C is the subspace generated by those elements which extend to either $L_x - \{\infty\}$ or $L_x - \{0\}$. The link between $H^1(L_x, \mathcal{O}(k))$ and the integral formulae is that the integral formulae give a specific calculation of the coefficients in these Laurent series (the usual formula in disguise). Laurent series enter into Corrigan's discussion of monopoles (in these proceedings) for much the same reason. The Penrose transform for massless fields now follows along exactly the same lines as the above discussion : For U an open subset of $\mathbb{M}$ and U' the corresponding open subset of $\mathbb{P}$, an element $f \in H^1(U', \mathcal{O}(-n-2))$ produces, by restriction to lines L_x, a section $\mathcal{P}f \in \Gamma(U, \mathcal{O}_{(A'B'...D')[E'F']})$ which satisfies the massless field equations of helicity $n/2$ for $n \geq 0$. Again the geometry of intersecting lines manifests itself in a differential equation (cf. the Radon transform [9]). Under mild topological assumptions on U it can be shown that $\mathcal{P}$ is an isomorphism [5]. One of the surprising aspects of the Penrose transform is that it works for negative helicity too i.e. for n negative $H^1(U', \mathcal{O}(-n-2))$ still represents fields of helicity $n/2$ (even though $H^1(L_x, \mathcal{O}(-n-2))$ always vanishes). The field can be constructed by restriction to the nth formal neighbourhood of L_x rather than L_x itself but the geometry of intersecting lines manifests itself more naturally as a "potential/gauge" description of the field. This turns out to be particularly effective when $n = -2$ so that $H^1(U', \mathcal{O})$ represents solutions of the anti-self-dual Maxwell equations as potential (1-form) modulo gauge (exact 1-forms). In this case Ward observed [27] that one can exponentiate this correspondence to obtain the gauge theory of anti-self-dual Maxwell fields as represented by line bundles on U'. He then showed that this generalizes to give a correspondence (the Ward correspondence) between anti-self-dual Yang-Mills fields and certain vector bundles on U'. This correspondence gives a systematic way of dealing with these non-linear equations but even in the linear case the twistor description is useful, for example :

<u>The Scalar Product</u> : A more usual way of solving the massless field equations is to take the Fourier transform. This splits a general solution on M into a sum of "positive frequency" and "negative frequency" parts. This splitting may be described twistorially as follows. The Hermitian form $\| \|^2$ on T divides $\mathbb{P}$ into three pieces $\mathbb{P}^+, P, \mathbb{P}^-$ according to whether $\| \|^2$ is positive, zero, or negative. Recall

that M may be identified as the space of lines in P . A cohomology class defined in a neighbourhood of P gives rise to a (real analytic) field on M . Splitting this field into positive and negative frequency parts corresponds to splitting the cohomology class (by means of the Mayer-Vietoris sequence) into a part which extends to a neighbourhood of $\overline{\mathbb{P}^+}$ and a part which extends to a neighbourhood of $\overline{\mathbb{P}^-}$. The classical scalar product is an Hermitian form on positive frequency fields or, equivalently, a bilinear form $\langle\phi|\psi\rangle$ for ϕ positive frequency of helicity $n/2$ and negative frequency of helicity $-n/2$ (since complex conjugation changes the sign of the frequency and helicity). For Maxwell fields, for example, it may be defined as follows. Suppose $F \in \Gamma(M,\Omega^2_+)$ is positive frequency and $G \in \Gamma(M,\Omega^2_-)$ is negative frequency (satisfying Maxwell's equations $dF = 0 = dG$) . Since $H^2(M,\mathbb{Z}) = 0$ we can choose $\phi \in \Gamma(M,\Omega')$ such that $d\Phi = G$ (ϕ is a potential for the field G). Now form $F\wedge\phi$. This is a closed 3-form ($F\wedge G = 0$ because F is self-dual, G anti-self-dual) and a different choice of Φ modifies $F\wedge\Phi$ by an exact 3-form. Thus if we form $\int_S F\wedge\Phi$ for a space-like hypersurface S , then the result is independent of choice of S and Φ . This is the scalar product $\langle F|G\rangle$. The definition for other helicities is similarly assymetric i.e. one of the fields is given a potential/gauge description. This indicates that a twistor description is natural and indeed this is the case. For $\phi \in H^1(\overline{\mathbb{P}^+},\mathcal{O}(-n-2))$ and $\psi \in H^1(\overline{\mathbb{P}^-},\mathcal{O}(n-2))$ first form their cup product $\phi \cup \psi \in H^2(P,\mathcal{O}(-4))$. The Mayer-Vietoris sequence for $\overline{\mathbb{P}^+}$ and $\overline{\mathbb{P}^-}$ gives an isomorphism $H^2(P,\mathcal{O}(-4)) \overset{\cong}{\to} H^3(\mathbb{P},\mathcal{O}(-4))$. Now use the canonical identification on $\mathbb{P}$, $\mathcal{O}(-4) = \wedge^4 T \otimes \Omega^3$. A choice of isomorphism $\wedge^4 T \simeq \mathbb{C}$ induces, therefore, an isomorphism $H^2(P,\mathcal{O}(-4)) \overset{\cong}{\to} H^3(\mathbb{P},\Omega^3) = \mathbb{C}$, this last identification coming from Serre duality [26]. In physics the scalar product is used to define the Hilbert space completion of the fields on M of a fixed helicity. This is the first step on the road to quantum mechanics. A mathematical aspect of the scalar product which is not so clear from the space-time point of view is that it is invariant under the conformal isometries of M . The corresponding statement for the twistor construction is that it is invariant under the action of SU(2,2) induced from the action of SU(2,2) on T preserving $\langle\ ,\ \rangle$. The invariance of the twistor construction is clear : the matrix having determinant 1 is reflected in the preservation of our chosen isomorphism $\wedge^4 T \simeq \mathbb{C}$. However, there is one problem with this twistor description. In order to define the scalar product as an Hermitian form on $H^1(\overline{\mathbb{P}^+},\mathcal{O}(-n-2))$ we need to have a complex conjugation $H^1(\overline{\mathbb{P}^+},\mathcal{O}(-n-2)) \to H^1(\overline{\mathbb{P}^-},\mathcal{O}(n-2))$ which is the usual complex conjugation on space-time. An integral formula for this transform (the twistor transform) was given by Penrose [21] and a cohomological reformulation was achieved by Ginsberg [10]. It enjoys many properties in common with the Fourier transform as explained in [7] . A remaining problem is to show, without resorting to arguments in Minkowski space, that $\langle\ |\ \rangle$ is a positive definite form on $H^1(\overline{\mathbb{P}^+},\mathcal{O}(-n-2))$. This would complete a rather natural construction of this unitary representation of SU(2,2) .

<u>Other Field Equations</u> : The twistor theory discussed so far is biased towards self-dual or anti-self-dual equations. In particular, such equations are source free. There is a more symmetric analogue of twistor theory which was used by Isenberg, Yasskin, and Green [15], and independently by Witten [31] to describe the full source free Yang-Mills equations. More recently Henkin and Manin have shown how the source may be interpreted. Again, the non-linear case is no more difficult than the linear case (Maxwell's equations). A brief description is as follows. Recall that there was a choice in our definition of twistor space. The other choice would lead to $\mathbb{P}^*$, the space of 3-dimensional subspaces of T. To combine these we define A as the space of (1,3)-flags in T (the line space of [15]). This 5-dimensional complex manifold (ambitwistor space) is a hypersurface in $\mathbb{P} \times \mathbb{P}^*$. For $x \in \mathbb{M}$ we obtain a quadric $L_x \times L_x^{\perp}$ in A. Conversely, A may be regarded as the space of complex null geodesics in $\mathbb{M}$. Maxwell's equations with source are $dF = 0$, $d * F = J$ where J is a given closed 3-form. If U is an open subset of $\mathbb{M}$ and U'' is the region in A swept out by the corresponding quadrics then it is a straightforward analogy with the usual twistor case to see that an element of $H^1(U'',\mathcal{O})$ gives rise to a solution of $dF = 0$. The question is to see how to impose $d * F = J$. The answer is in terms of Griffiths obstructions [11] to extending the cohomology class to the formal neighbourhoods of U'' in $\mathbb{P} \times \mathbb{P}^*$. Up to second order there is nothing gained or lost : $H^1(U'',\mathcal{O}_{(2)}) \overset{\cong}{\to} H^1(U'',\mathcal{O})$. The proof of this is implicit in the power series calculations of Isenberg et al. and more explicitly demonstrated by Pool [25]. More recently, Buchdahl [4] has given a much more invariant proof. From the exact sequence $0 \to \mathcal{O}(-3,-3) \to \mathcal{O}_{(3)} \to \mathcal{O}_{(2)} \to 0$ we obtain an obstruction in $H^2(U'',\mathcal{O}(-3,-3))$ to the extension to third order. The Penrose transform identifies $H^2(U'',\mathcal{O}(-3,-3))$ as the closed 3-forms on U. This obstruction is the current J. The space A is useful for other equations :

<u>The massive Dirac equations</u> : Since twistor theory is intrinsically conformally invariant the introduction of mass has proved somewhat problematic (bricks do not travel at the speed of light). There are at least three possible approaches :

a) There are Penrose integral formulae [23] based on A but the integrand is required to satisfy an auxillary equation. This method has been investigated by Hodges [13] who shows how to construct suitable integrands. All solutions of the Dirac equations can be described in this way [6].

b) Using the language of obstructions as above Henkin and Manin [12] have shown how to translate the Dirac equations word for word to obtain an equivalent statement regarding cohomology classes external to the cohomology groups (one can also couple these equations to a Yang-Mills background by tensoring the coefficients with the corresponding Ward bundle ← a general comment which applies to all the equations we have discussed so far). This dictionary can be used for other equations too (e.g. ϕ^4 (Buchdahl)).

c) LeBrun has shown [16] how to use the obstruction theory to give an internal description of the Klein-Gordan equation on space-time as a cohomology group on the first formal neighbourhood on A in $\mathbb{P} \times \mathbb{P}^*$. This is a very geometric procedure which works for a curved space-time too.

References.

[1] M.F. Atiyah and R.S. Ward, Instantons and algebraic geometry, Comm. Math. Phys. 55, 111-124 (1977).

[2] M.F. Atiyah, N.J. Hitchin, V.G. Drinfeld, and Yu.I. Manin, Construction of instantons, Phys. Lett. 65A, 185-187 (1978).

[3] M.F. Atiyah, Geometry of Yang-Mills fields, Lezioni Fermiane, Scuola Normale Superiore, Pisa 1979.

[4] N.P. Buchdahl, to appear.

[5] M.G. Eastwood, R. Penrose, and R.O. Wells, Jr., Cohomology and massless fields, Comm. Math. Phys. 78, 305-351 (1981).

[6] M.G. Eastwood, On the twistor description of massive fields, Proc. R. Soc. Lond. A374, 431-445 (1981).

[7] M.G. Eastwood and M.L. Ginsberg, Duality in twistor theory, Duke Math. J. 48, 177-196 (1981).

[8] H.D. Fegan, Conformally invariant first order differential operators, Quart J. Math. Oxford (2) 27, 371-378 (1976).

[9] I.M. Gel'fand, M.I. Graev, and N.Ya. Vilenkin, Integral geometry and representation theory : Generalized functions volume 5, Academic Press 1966.

[10] M.L. Ginsberg, A cohomological scalar product construction, [14] 293-300.

[11] P.A. Griffiths, The extension problem in complex analysis II : embeddings with positive normal bundle, Amer. J. Math. 88, 366-446 (1966).

[12] G.M. Henkin and Yu.I. Manin, Twistor description of classical Yang-Mills-Dirac fields, Phys. Lett. 95B, 405-408 (1980).

[13] A.P. Hodges, The description of mass in the theory of twistors, Ph.D. thesis, London 1975.

[14] L.P. Hughston and R.S. Ward (eds.), Advances in twistor theory, Research Notes in Math. 37, Pitman 1979.

[15] J. Isenberg and P.B. Yasskin, Twistor description of non-self-dual Yang-Mills fields, [17] 180-206.

[16] C.R. LeBrun, The first formal neighbourhood of ambitwistor space for curved space-time, to appear.

[17] D.E. Lerner and P.D. Sommers (eds.), Complex manifold techniques in theoretical physics, Research Notes in Math. 32, Pitman 1979.

[18] J. Milnor and J.D. Stasheff, Characteristic classes, Princeton University Press 1974.

[19] R. Penrose, Twistor algebra, J. Math. Phys. 8, 345-366 (1967).

[20] R. Penrose, Solutions of the zero-rest-mass equations, J. Math. Phys. 10, 38-39 (1969).

[21] R. Penrose and M.A.H. Mac Callum, Twistor theory : and approach to the quantisation of fields and space-time, Phys. Rep. 6C 241-316 (1972).

[22] R. Penrose, Twistors and particles : an outline, In : Quantum theory and the structure of space-time (eds. L. Castell, M. Drieschner, C.F. von Weizsäcker), Munich Verlag 1975.

[23] R. Penrose, Twistor theory, its aims and achievements, In : Quantum gravity : an Oxford symposium, 268-407, Clarendon Press 1975.

[24] R. Penrose, On the twistor description of massless fields, [17] 55-91.

[25] R. Pool, Ph.D. thesis, Rice University 1981.

[26] J.-P. Serre, Un théorème de dualité, Comm. Math. Helv. 29, 9-26 (1955).

[27] R.S. Ward, The twisted photon : massless fields as bundles, [14] 132-135.

[28] R.S. Ward, On self-dual gauge fields, Phys. Lett. 61A, 81-82 (1977).

[29] R.S. Ward, A Yang-Mills-Higgs monopole of charge 2, Comm. Math. Phys. 79, 317-325 (1981).

[30] R.O. Wells, Jr., Complex manifolds and mathematical physics, Bull. Amer. Math. Soc. 1, 296-336 (1979).

[31] E. Witten, An interpretation of classical Yang-Mills theory, Phys. Lett. 77B, 394-398 (1978).

TWISTOR TECHNIQUES IN GAUGE THEORIES

R.S. Ward
Department of Mathematics
Trinity College
Dublin, Ireland

I. Introduction.

These lectures are about the so-called twistor method for dealing with certain non-linear pdes that crop up in theoretical physics. The previous conference in this series dealt mainly with equations such as Korteweg-de-Vries and Sine-Gordon, and with techniques such as Bäcklund transformations for solving them. By contrast, twistor theory applies to gauge theories (i.e. geometric theories) in four-dimensional space-time.

In order to motivate the subsequent discussion, let me begin by talking about a simple linear equation, namely the wave equation $\Box\phi = 0$ in flat space-time. The "standard" way of generating solutions of this equation is to use the Fourier transform: when one transforms the problem from coordinate space into momentum space, the differential equation disappears, and so one can write down the general solution of the equation as an integral (the inverse Fourier transform). The limitation of this "Fourier" approach is that it does not, apparently, generalize to non-linear equations.

There are, however, other ways of solving $\Box\phi = 0$. One such is to take a complex-analytic function F of three complex variables, and to put

$$\phi(x,y,z,t) = \oint F[z + t + (x+iy)\zeta,\ x - iy + (z-t)\zeta,\ \zeta]\ d\zeta\ , \tag{1}$$

the integral being taken, for fixed xyzt, over any contour in the complex ζ-plane. The field ϕ defined by (1) is automatically a solution of $\Box\phi = 0$; furthermore, <u>every</u> real-analytic solution can be obtained in this way.

This sort of formula has been known for a long time [1], and has been extensively studied in recent years [2]. The remarkable thing is that, unlike the Fourier formula, it generalizes to certain non-linear equations such as the self-dual Yang-Mills and self-dual Einstein equations. This generalization, and some of its applications, are what will be described below.

II. Self-Dual Gauge Fields and the Twistor Construction.

The self-dual Yang-Mills equations are rather remarkable equations that have turned up in several different places. Let me describe briefly what they are. Let A_μ (for each value of $\mu = 0,1,2,3$) be an n x n matrix of complex-valued functions on flat space-time. This A_μ is called the gauge potential (or connection 1-form). From it one can compute the gauge field (or curvature):

$$F_{\mu\nu} = \partial_\mu A_\nu - \partial_\nu A_\mu + i[A_\mu, A_\nu],$$

where ∂_μ denotes the partial derivative with respect to x^μ, the standard linear co-ordinates on Minkowski space-time. The self-duality equations are

$$*F_{\mu\nu} \equiv \tfrac{1}{2}\varepsilon_{\mu\nu\rho\sigma} F^{\rho\sigma} = i \; F_{\mu\nu}, \qquad (2)$$

where $\varepsilon_{\mu\nu\rho\sigma}$ is the usual alternating tensor. The system (2), considered as equations for A_μ, is a quasi-linear hyperbolic system of coupled first-order pdes.

Let us turn now to the twistor construction. The basic twistor geometry has been described in Mike Eastwood's lectures. For our purposes, it is sufficient to depict the geometry as in figure 1. Here R is a region of flat space-time, and T

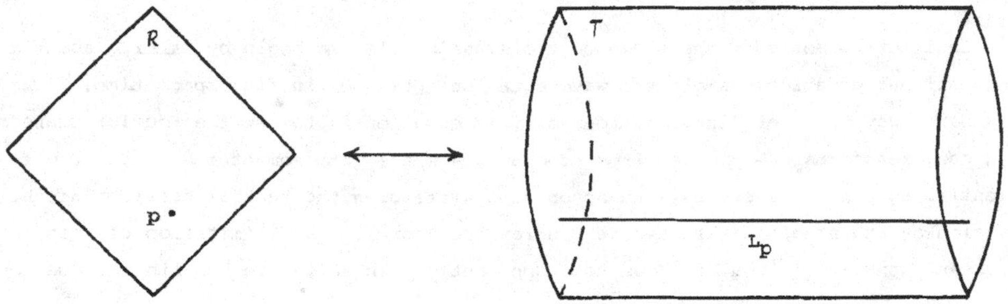

Figure 1

is the corresponding region in projective twistor space; thus T is an open subset of complex projective 3-space $P_3(C)$, a 3-dimensional complex manifold. The points p in space-time correspond to complex projective lines L_p in twistor space; each L_p is intrinsically a sphere S^2. Of course, R may consist of the whole of space-time, or even compactified space-time; I have chosen a general region R in order to emphasize the local nature of the theorem that follows.

Theorem. There is a natural one-to-one correspondence between

(a) solutions of the self-duality equations (2) in R; and

(b) complex-analytic vector bundles E over T, of rank n, such that E restricted to L_p is trivial for all $p \in R$.

Remarks.

(i) The gauge group is GL(n,C) (in other words, A_μ and $F_{\mu\nu}$ are n x n matrices which do not, in general, satisfy any algebraic conditions). There are other versions of the theorem which allow for different gauge groups, such as SU(n), and for Euclidean 4-space rather than Minkowski space-time [3].

(ii) I shall not explain what the words in part (b) of the above theorem mean. The crucial point is that there is no differential equation in (b): the non-linear pde has been "transformed away". I want now to indicate how one can use this fact to construct solutions.

III. Non-Abelian Magnetic Monopoles.

To illustrate the power of the twistor method, I shall describe how it may be applied to the problem of non-abelian magnetic monopoles. Ed Corrigan has already told us how a special case of the self-duality equations (2) may be interpreted as the equations which describe the simplest type of non-abelian magnetic monopole. Essentially, we impose the conditions

(a) the A_μ are independent of x^o and smooth in x^1, x^2, x^3;

(b) the A_μ are tracefree 2 x 2 matrices; A_o is anti-hermitian, and A_1, A_2 and A_3 are hermitian;

(c) the hermitian matrix $\Phi = i A_o$ satisfies

$$\mathrm{tr}\ \Phi^2 = 2 - n/r + O(r^{-2}) \tag{3}$$

as $r \to \infty$, where $r^2 = (x^1)^2 + (x^2)^2 + (x^3)^2$ and n is some real number.

Remarks.

(i) The pair of fields (A_j, Φ) is called an SU(2) magnetic monopole. (j,k, ... run over 1,2,3.)

(ii) The self-duality equations (2) imply the so-called Bogomolny equations

$$F_{jk} = - \varepsilon_{jk\ell}\ D_\ell\ \Phi, \tag{4}$$

where $F_{jk} = \partial_j A_k - \partial_k A_j + i[A_j, A_k]$ and $D_j \Phi = \partial_j \Phi + i [A_j, \Phi]$.

(iii) The number n appearing in equation (3) must be a non-negative integer. This is not obvious, but is true for topological reasons. This integer is called the <u>topological charge</u>.

(iv) It has been proved that for each n = 1,2,3, ... there exists a (4n-1)-parameter family of monopoles [4]. However, this existence proof is non-constructive, and the problem remained of finding out what the monopole solutions look like. Only for n = 1 was an explicit solution known [5]; this is a spherically symmetric monopole which was found because of its high degree of symmetry.

Let us move on now to consider how the twistor construction procedure of section II may be applied to this particular problem. One starts with a 2 x 2 matrix $g(\gamma,\zeta)$ of complex-analytic functions of the two complex variables γ and ζ, defined for all γ and for ζ in some neighbourhood of $|\gamma| = 1$, and satisfying

$$\det(g) = 1$$
$$g(\bar{\gamma}, -\bar{\zeta}^{-1}) = g(\gamma,\zeta)^*, \tag{5}$$

where the * on the right-hand side of (5) denotes conjugate transpose. In fact, g is the transition matrix which determines the vector bundle E appearing in the theorem of section II; and therefore, according to the theorem, g determines A_μ. This works as follows. First, substitute $\gamma = \gamma(x^j,\zeta) \equiv (x^1 + ix^2)\zeta - 2x^3 - (x^1 - ix^2)\zeta^{-1}$ into $g(\gamma,\zeta)$ and "split" g:

$$g(\gamma(x^j,\zeta),\ \zeta) = \hat{h}(x^j,\zeta)\, h(x^j,\zeta)^{-1}, \tag{6}$$

where h and $\hat{h}$ are 2 x 2 matrices with unit determinant, and are complex-analytic in ζ for $|\zeta| \leq 1$ and $|\zeta| \geq 1$ (including ∞) respectively. Then the matrices A_μ are given by simple formulae in terms of h and $\hat{h}$. For example,

$$\Phi = iA_0 = \tfrac{1}{2}h_0^{-1}\,\partial_3\, h_0 - \tfrac{1}{2}\hat{h}_\infty^{-1}\,\partial_3\,\hat{h}_\infty ,$$

where $h_0(x^j) = h(x^j,0)$ and $\hat{h}_\infty(x^j) = \hat{h}(x^j,\infty)$. The formulae for A_1, A_2 and A_3 are similar.

Remarks.

(i) The fields obtained in this way are automatically solutions of the equations (4). Furthermore, every solution can be obtained by this method. This follows from the theorem of section II.

(ii) The smoothness and hermiticity requirements mentioned earlier in this section are also automatically satisfied, although the boundary condition (3) is not: it still has to be imposed.

(iii) That the splitting (6) should be possible, imposes a condition on the matrix g. I shall say more about this below.

(iv) The splitting is the non-linear generalization of "Taylor-Laurent" splitting in complex analysis.

Indeed, if g were a 1 x 1 matrix, h and $\hat{h}$ would be given by formulae like $h = \exp\oint \log g$, where $\oint$ is a Cauchy integral. In this connection, observe that

$$\phi(x^j) = \oint f(\gamma(x^j,\zeta),\ \zeta)\, d\zeta,$$

where f is any analytic function of two complex variables, is the general solution of the three-dimensional Laplace equation $\nabla^2\phi = 0$ [1].

The difficult part of the above solution procedure (there has to be a catch somewhere!) lies in carrying out the splitting (6) explicitly. In general, no algorithm is known which achieves it. There is, however, a class of matrices g which can be split in a fairly explicit way, namely those which are upper triangular:

$$\tilde{g} = \begin{bmatrix} L & \Gamma \\ 0 & L^{-1} \end{bmatrix}. \tag{7}$$

The problem with this is that such matrices cannot satisfy the reality condition (5), unless they are in fact diagonal matrices, which would be far too restrictive. But we are saved by the fact that the mapping $g \to (\Phi, A_j)$ is not one-to-one: there is some freedom in g, namely

$$g \mapsto \hat{\lambda}\, g\, \lambda , \tag{8}$$

where λ and $\hat{\lambda}$ are 2 x 2 matrices of complex-analytic functions of γ and ζ. λ must be defined for $|\zeta| \leq 1$ and for all γ including $\gamma = \zeta^{-1}$, while $\hat{\lambda}$ must be defined for $|\zeta| \geq 1$ and all γ including $\gamma = \zeta$. Geometrically, the transformation (8) corresponds to changing coordinates in the vector bundle E, and such a change does not affect Φ or A_j.

To sum up: the general matrix g gives, in principle, all solutions of the non-linear pde. In practice, the calculations are tractable if g can be "upper-triangularized" by a transformation (8). Fortunately, a recently established theorem states that this is good enough for the monopole problem: all monopoles of topological charge n can be obtained from matrices of the form (7), where $L = \zeta^n e^\gamma$ [6]. For these upper triangular matrices the splitting of g boils down to expanding Γ in a Taylor-Laurent series and then doing some algebra, and the resulting structure may be described as a sequence of ansätze A_1, A_2, ... which convertlinear fields into solutions of the non-linear equations [7]. This is described more fully in Ed Corrigan's lectures in this volume.

The remaining problem is to find those functions Γ which give us monopole solutions with the desired properties. Work is still in progress on this, although considerable headway has already been made, and it looks as if the problem is almost solved [8]. The most difficult part (returning to a remark made earlier) is to prove that the splitting (6) is possible; it amounts to a condition on the function Γ, but this condition is rather hard to handle.

IV. Other Applications of Twistor Methods,

In the previous section I gave an outline of how twistor theory may be applied to one particular non-linear problem. I should like to conclude by mentioning a few other applications.

The basic theorem of section II is useful not only for SU(2) monopoles, but can be applied to problems involving different gauge groups (such as SU(3)) and different global/boundary conditions (for example Yang-Mills theory on a 4-torus [9] or a 4-sphere [3,10]). Most of these topics are still at an early stage of development, although one has been around for several years: that of Yang-Mills fields on the 4-sphere S^4, which are called "instantons".

Instantons arise in the "path integral" approach to the quantization of gauge theories, where for technical reasons one works in Euclidean 4-space rather than space-time (i.e. signature ++++ rather than +---) [10]. One is interested in Yang-Mills connections A_μ on 4-space which have finite action

$$S(A) = \tfrac{1}{2}\int \mathrm{tr}\, F \wedge {*F}, \qquad (9)$$

where F is the curvature of A and *F is the dual of F (cf. equation 2). In particular one wants to find the stationary points of the action functional (9), and these are what are called instantons.

The condition $S(A) < \infty$ implies that the gauge field extends from Euclidean 4-space to its conformal compactification S^4 (note that the functional (9) is conformally invariant). So the problem becomes one of looking for critical points of (9) on S^4. As Prof. Friedrich has described in his lectures, the local minima of (9) are self-dual (or anti-self-dual). It is an important unsolved problem as to whether there are any other critical points of the action. But in any event the self-dual (and the anti-self-dual) instantons can be constructed by twistor methods. Indeed, self-dual instantons correspond to analytic vector bundles of rank 2 over $P_3(C)$, satisfying a few additional conditions [11]. By coincidence, such bundles have been extensively studied in recent years by algebraic geometers. They came up with two ways of constructing bundles: the first is to use upper triangular matrices analogous to those mentioned in section II [11], and the second is to use non-linear matrix algebra [12]. Each of these methods simplifies the original problem considerably, but it still remains very complicated.

The final topic I should like to mention concerns the self-dual Einstein equations. These have cropped up in several different places, one of the most important of which is the path-integral approach to quantum gravity; cf. [10].

Let g be a 4-dimensional Riemannian matric. Its curvature tensor $\underline{\text{Riem}}$ decomposes into four irreducible pieces:

$$\underline{\text{Riem}} = \underline{C}^+ + \underline{C}^- + \underline{\text{Ric}} + \underline{R},$$

namely $\underline{C}^+$ (the self-part of the conformal curvature tensor), $\underline{C}^-$ (the anti-self-part), $\underline{\text{Ric}}$ (the Ricci tensor), and $\underline{R}$ (the scalar curvature). Einstein's equations (with cosmological constant) are

$$\underline{\text{Ric}} = \frac{1}{4}\underline{R}\, g\,; \tag{10}$$

in order words, the trace-free part of the Ricci tensor vanishes. Twistor theory provides a way of constructing self-dual solutions of (10), i.e. solutions which have $\underline{C}^- = 0$. Instead of building a bundle over twistor space $\mathcal{T}$ (as one does for the Yang-Mills problem), one deforms the twistor space itself; the non-linear pde (10) is converted into a problem concerning deformations of complex manifolds [13]. This correspondence can, and has been, used to construct new solutions of Einstein's equations [14]. Its full potential still remains to be exploited.

References.

1. E.T. Whittaker, A Course of Modern Analysis (Cambridge University Press, 1902)
   H. Bateman, Proc. Lond. Math. Soc. (2), I (1904), 451-458.

2. M.G. Eastwood, R. Penrose and R.O. Wells, Jr., Commun. Math. Phys. 78 (1981), 305-351.

3. M.F. Atiyah, N.J. Hitchin and I.M. Singer, Proc. Roy. Soc. Lond. A362 (1978), 425-461.

M.F. Atiyah, Geometry of Yang-Mills Fields (Scuola Normale Superiore, Pisa, 1979).

4. E.J. Weinberg, Phys. Rev. D20 (1979), 936-944.
A. Jaffe and C.H. Taubes, Vortices and Monopoles (Birkhauser, Boston, 1980).

5. M.K. Prasad and C.M. Sommerfield, Phys. Rev. Lett. 35 (1975), 760-762.

6. N.J. Hitchin, to be published.

7. E.F. Corrigan, D.B. Fairlie, R.G. Yates and P. Goddard, Commun. Math. Phys. 58 (1978), 223-240.

R.S. Ward, Commun. Math. Phys. 80 (1981), 563-574.

8. R.S. Ward, Commun. Math. Phys. 79 (1981), 317-325.
M.K. Prasad, Commun. Math. Phys. 80 (1981), 137-149.
R.S. Ward, Phys. Lett. B 102 (1981), 136-138.
E.F. Corrigan and P. Goddard, Commun. Math. Phys. 80 (1981), 575-587.

9. G.'tHooft, Commun. Math. Phys. 81 (1981), 267-275.

10. T. Eguchi, P.B. Gilkey and A.J. Hanson, Phys. Repts 66 (1980), 213-393.

11. M.F. Atiyah and R.S. Ward, Commun. Math. Phys. 55 (1977), 117-124.

12. M.F. Atiyah, N.J. Hitchin, V.G. Drinfeld and Yu. I. Manin, Phys. Lett. A65 (1978), 185-187.

13. R. Penrose, Gen. Rel. Grav. 7 (1976), 31-52.
R.S. Ward, Commun. Math. Phys. 78 (1980), 1-17.

14. N.J. Hitchin, Math. Proc. Camb. Phil. Soc. 85 (1979), 465-476.

SIMPLE PSEUDOPOTENTIALS FOR THE Kd V-EQUATION

Pierre MOLINO

Differentiability is assumed to be real analytic.

Let $M = \mathbb{R}^5$, endowed with coordinates (x,t,u,z,p). We consider on M the exterior differential system [EDS] in the sense of E. Cartan [1]

$$(1)\qquad \begin{cases} \alpha \equiv du \wedge dt - z\, dx \wedge dt = 0 \\ \beta \equiv dz \wedge dt - p\, dx \wedge dt = 0 \\ \gamma \equiv -du \wedge dx + dp \wedge dt + 12\, uz\, dx \wedge dt = 0 \end{cases}$$

which is closed in the sense that, if $\mathcal{J}$ is the ideal of differential forms on M generated by α, β, γ, then we have $d\alpha, d\beta, d\gamma \in \mathcal{J}$.

We denote by $\pi : M \to \mathbb{R}^2$ the projection $\pi(x,t,u,z,p) = (x,t)$.

A submanifold S of M is an integral manifold of (1) if the induced forms $\alpha_S, \beta_S, \gamma_S$ are 0. Let $s : \mathbb{R}^2 \to M$ be a section of π, defined by

$$s(x,t) = (x,t,u(x,t),z(x,t),p(x,t)).$$

Estabrook-Wahlquist [2] observed that $S = s(\mathbb{R}^2)$ is an integral manifold of (1) iff $z = u_x$ and $p = u_{xx}$, where $u(x,t)$ is a solution of the KdV-equation

$$(2)\qquad u_t + u_{xxx} + 12\, u\, u_x = 0$$

Estabrook-Wahlquist used this representation of (2) by the EDS (1) in order to obtain pseudopotentials for the kd V-equation. We will introduce in §I this notion of pseudopotential in a slightly generalized form.

Using some restrictive hypothesis [like space-time independance] they obtained an explicit method to calculate such pseudopotentials.

Our purpose is to given a global classification [up to a natural isomorphism relation] of all the [simple] pseudopotentials for the KdV-equation, without any restrictive hypothesis.

We obtain two different types of models : first, potential models, some of which depend on space and time. On the other late, a unique model which is not a potential : essentially the space-time independant pseudopotential discovered by Estabrook-Wahlquist.

This unexpected unicity property gives, in a certain sense, an a posteriori justification of Estabrook-Wahlquist's argument.

I - ADAPTED CONNECTIONS AND SIMPLE GENERALIZED PSEUDOPOTENTIALS.

Let $\tilde{\pi} : \tilde{M} \to M$ be a locally trivial fibration with F as typical fiber. A Cartan-Ehresmann connection on $(\tilde{M},\tilde{\pi},M)$ is a field $\mathcal{H}$ of contact elements on $\tilde{M}$ which is supplementary of the field $\mathcal{V}$ of $\tilde{\pi}$-vertical elements. If $\tilde{m} \in \tilde{M}$, $\mathcal{H}_{\tilde{m}}$ is the horizontal element at $\tilde{m}$ of the connection.

In order to describe the theory of simple pseudopotentials, we will consider in this paper the case where $F = \mathbb{R}$. If one takes $\mathbb{R}^p$ as typical fiber, he would obtain the theory of multiple pseudopotentials.

Let

$$\Phi_U : \tilde{U} = (\tilde{\pi})^{-1}(U) \to U \times \mathbb{R} \tag{3}$$

be a local trivialisation of $(\tilde{M},\tilde{\pi},M)$ in the open subset U of M. Φ_U defines coordinates (x,t,u,z,p,y) in $\tilde{U}$. There exists in $\tilde{U}$ a unique connection form

$$\omega_U = dy - Adx - Bdt - Cdu - Ddz - Edp \tag{4}$$

such that $\omega_U = 0$ defines the connection $\mathcal{H}$. A,B,C,D,E are functions on $\tilde{U}$.

Now, let $\tilde{\mathcal{J}}$ be the ideal of differential forms on $\tilde{M}$ generated by $\tilde{\pi}^*\alpha, \tilde{\pi}^*\beta, \tilde{\pi}^*\gamma$ and those 1-forms which are 0 in restriction to $\mathcal{H}$. This ideal defines a prolonged EDS on $\tilde{M}$. In $\tilde{U}$, this prolonged EDS is defined by

$$\begin{cases} \tilde{\pi}^*\alpha = 0 \\ \tilde{\pi}^*\beta = 0 \\ \tilde{\pi}^*\gamma = 0 \\ \omega_U = 0 \end{cases} \tag{5}$$

The connection $\mathcal{H}$ will be said adapted to (1) if the prolonged EDS is closed. Locally, this closure condition is equivalent to

$$d\omega_U = 0 \qquad \text{modulo} \qquad \tilde{\pi}^*\alpha,\ \tilde{\pi}^*\beta,\ \tilde{\pi}^*\gamma,\ \omega_U. \tag{6}$$

The geometrical interpretation of this property is the following one : let S be an integral manifold of (1). Then $\mathcal{H}$ induces in $(\tilde{\pi})^{-1}(S) = S \times \mathbb{R}$ an integrable connection whose maximal integral manifolds are the horizontal lifts of S. Such an horizontal lift $\tilde{S}$ is an integral manifold of the prolonged EDS such that $S = \tilde{\pi}(\tilde{S})$. Conversely, if $\tilde{S}$ is an integral manifold of the prolonged EDS, then $\tilde{\pi}(\tilde{S})$ is an integral manifold of (1). In other words, the initial EDS and the prolonged system are essentially equivalent.

In coordinates (x,t,u,z,p,y) associated to the local trivialisation (3), the horizontal lifts of an integral manifold S are determined by the condition

$$(7) \qquad dy = Adx + Bdt + Cdu + Ddz + Edp.$$

Functions y on S satisfying the differential equation (7) will be referred to as generalized simple pseudopotentials for the EDS (1), associated to the adapted connection $\mathcal{H}$.

If $u(x,t)$ is a solution of the Kd V-equation, putting on (7)

$$u = u(x,t) \ , \quad z = u_x(x,t) \quad , \quad p = u_{xx}(x,t)$$

we will obtain an integrable differential equation in the unknown function $y(x,t)$. Solutions of this equation [depending on the considered solution $u(x\ t)$ of (2)] are generalized simple pseudopotentials for the KdV-equation.

REMARK - In the original paper by Estabrook-Wahlquist, the fiber bundle $(\tilde{M},\tilde{\pi},M)$ is assumed to be the trivial bundle $M \times \mathbb{R}$, the pseudopotential equation (7) having the particular form

$$(8) \qquad dy = Adx + Bdt$$

This form is related to the fact that x,t are considered as independant variables. However, if we are looking at the system (1), it seems to be more natural to consider the generalized form (7) of the pseudopotential equation.

II - FOLIATED TRIVIALISATIONS OF $(\tilde{M},\tilde{\pi},M)$.

From now on, we consider a Cartan-Ehresmann connection $\mathcal{H}$ on $(\tilde{M},\tilde{\pi},M)$ which is adapted to (1). Our idea is to use particular trivialisations of $(\tilde{M},\tilde{\pi},M)$ such that the corresponding pseudopotential equation (7) has a simplified form.

If $\Phi_U, \Phi'_U : \tilde{U} \to U \times \mathbb{R}$ are two different trivialisations in the open subset U of M, we consider the corresponding coordinates systems (x,t,u,z,p,y) and (x,t,u,z,p,z'). The change of coordinates is determinated by

$$(9) \qquad y' = \varphi(x,t,u,z,p,y), \text{ with } \varphi_y \neq 0.$$

If the connection forms have the following expressions :

$$(10) \qquad \begin{cases} \omega_U = dy - Adx - Bdt - Cdu - Ddz - Edp & \text{in the trivialisation } \Phi_U \\ \omega'_U = dy' - A'dx - B'dt - C'du - D'dz - E'dp & \text{in the trivialisation } \Phi'_U \end{cases}$$

then we have

$$(11)\qquad \begin{cases} A'(x,t,u,z,p,y') = \varphi_y\ A(x,t,u,z,p,y) + \varphi_x \\ B'(x,t,u,z,p,y') = \varphi_y\ B(x,t,u,z,p,y) + \varphi_t \\ \text{etc...} \end{cases}$$

The first simplification of the generalized pseudopotential equation (7) will be obtained in the following way : let us observe that equations

$$x = x_o \quad , \quad t = t_o$$

define an integral manifold S_{x_o,t_o} of (1). Thus, $\mathcal{H}$ induces on $(\tilde{\pi})^{-1}(S_{x_o,t_o}) = S_{x_o,t_o} \times R$ an integrable connection. Moreover, as S_{x_o,t_o} is simply connected, the horizontal lifts of S_{x_o,t_o} are sections of $S_{x_o,t_o} \times \mathbb{R}$.

Using the terminology of foliations, we will consider the fibers of the fibration $\pi : M \to \mathbb{R}^2$ as leaves of a foliation $\mathfrak{F}_\pi$ on M. By the previous observation, we have in $\tilde{M}$ an horizontal lifted foliation $\tilde{\mathfrak{F}}_\pi$. The local trivialisation Φ_U of $(\tilde{M},\tilde{\pi},M)$ is a foliated trivialisation iff the leaves of $\tilde{\mathfrak{F}}_\pi$ have the following equations in local coordinates (x,t,u,z,p,y) :

$$x = x_o \quad , \quad t = t_o \quad , \; y = y_o$$

Such foliated trivialisations exist in a neighbourood of each point of M. With respect to such a trivialisation, the connection form has the following expression :

$$(12)\qquad \omega_U = dy - Adx - Bdt$$

In other words, with respect to foliated local trivialisations of $(\tilde{M},\tilde{\pi},M)$, the pseudopotential equation has the particular form (8) introduced by Estabrook-Wahlquist.

Now, if Φ'_U is another foliated trivialisation, the transition function (9) has the particular form

$$(13)\qquad y' = \varphi(x,t,y)$$

III - THE CLOSURE CONDITION.

Following Estabrook-Wahlquist, we will explicit the closure condition (6). We obtain

$$(14)\quad \begin{cases} A_z = A_p = 0 \\ B_p = -A_u \\ AB_y - BA_y + B_x - A_t + B_u z + B_z p + 12uzA_u = 0 \end{cases}$$

It is more convenient to present these relations in a slightly different way : we will introduce in $\widetilde{U}$ the $\widetilde{\pi}$-vertical vector fields :

$$(15)\quad \widetilde{A} = A\frac{\partial}{\partial y} \quad , \quad \widetilde{B} = B\frac{\partial}{\partial y}$$

(14) becomes

$$(16)\quad \begin{cases} \widetilde{A}_z = \widetilde{A}_p = 0 \; ; \; \widetilde{B}_p = -\widetilde{A}_u \\ [\widetilde{A},\widetilde{B}] + \widetilde{B}_x - \widetilde{A}_p + z\,\widetilde{B}_u + p\,\widetilde{B}_z + 12uz\,\widetilde{A}_u = 0 \end{cases}$$

From (16), we deduce

$$(17)\quad \widetilde{B} = -p\widetilde{A}_u + \frac{z^2}{2}\widetilde{A}_{uu} + z(\widetilde{A}_{ux} + [\widetilde{A},\widetilde{A}_u]) + \widetilde{B}^*$$

where the components of $\widetilde{A}$ and the components of $\widetilde{B}^*$ are functions of (x,t,u,z) only, with conditions

$$(18)_a\quad \widetilde{A}_{uuu} = 0$$

$$(18)_b\quad \widetilde{A}_{uux} + [\widetilde{A},\widetilde{A}_{uu}] = 0$$

$$(18)_c\quad \widetilde{B}^*_u = -12u\,\widetilde{A}_u - [\widetilde{A},\widetilde{A}_u]_x - [\widetilde{A},\widetilde{A}_{ux}] - \widetilde{A}_{uxx} - [\widetilde{A},[\widetilde{A},\widetilde{A}_u]])$$

$$(18)_d\quad [\widetilde{A},\widetilde{B}^*] + \widetilde{B}^*_x - \widetilde{A}_t = 0$$

Condition $(18)_a$ implies

$$(19)\quad \boxed{\widetilde{A} = 2\widetilde{X}_1 + 2u\,\widetilde{X}_2 + 3u^2\widetilde{X}_3}$$

where $\widetilde{X}_1, \widetilde{X}_2, \widetilde{X}_3$ are functions of (x,t,y). Coefficients in (19) are introduced in accordance with notations of [2].

Now, if Φ'_U is another foliated trivialisation, the new connection form is

$$(20)\quad \omega'_U = dy' - A'dx - B'dt$$

and, if $\widetilde{A}' = A'\frac{\partial}{\partial y'}$, $\widetilde{B}' = B'\frac{\partial}{\partial y'}$, we have by (11) and (13)

$$(21)\quad \widetilde{A}' = 2\widetilde{X}'_1 + 2u\,\widetilde{X}'_2 + 3u^2\widetilde{X}'_3$$

where

$$(22)\quad \begin{cases} \tilde{X}'_1 = \tilde{X}_1 + \frac{1}{2}\frac{\varphi_x}{\varphi_y}\frac{\partial}{\partial y} \\ \tilde{X}'_2 = \tilde{X}_2 \\ \tilde{X}'_3 = \tilde{X}_3 \end{cases}$$

$y' = \varphi(x,t,y)$ being the transition function.

The first important observation is that the $\tilde{\pi}$-vertical vector fields $\tilde{X}_2$ and $\tilde{X}_3$ are intrinsic ; thus, they are global analytic vector fields on $\tilde{M}$.

On the other late, if $\tilde{X}_1 = X_1 \frac{\partial}{\partial y}$, the differential equation

$$(23)\qquad 2X_1\varphi_y + \varphi_x = 0$$

where $\varphi(x,t,y)$ is an unknown function, has analytic local solutions such that $\varphi_y \neq 0$. Hence, it is always possible to use local coordinates (x,t,u,z,p,y) such that, with respect to these coordinates [which correspond to a local foliated trivialisation whose domain $\tilde{U}$ is not the preimage by $\tilde{\pi}$ of its projection $U = \tilde{\pi}(\tilde{U})$], we have $\tilde{X}_1 = 0$.

IV - GLOBAL MODELS IF $\tilde{X}_3 \equiv 0$.

In this case, analycity of $\tilde{X}_3$ implies that

$$\tilde{\Omega} = \{\tilde{m} \in \tilde{M} \;/\; \tilde{X}_{3\tilde{m}} \neq 0\}$$

is an open dense set in $\tilde{M}$.

If $\tilde{m} \in \tilde{\Omega}$, it is possible to change locally coordinates by a transition function $y' = \varphi(x,t,y)$ in such a way that $\tilde{X}_3 = \frac{\partial}{\partial y'}$. So, we can assume that in an open neighbourhood $\tilde{U}$ of $\tilde{m}$, we have coordinates such that $\tilde{X}_3 = \frac{\partial}{\partial y}$. In this case, $\tilde{A}_{uux} = 0$ and $(18)_b$ implies

$[\tilde{X}_1, \frac{\partial}{\partial y}] = [\tilde{X}_2, \frac{\partial}{\partial y}] = 0$. Thus $\tilde{X}_1 = X_1 \frac{\partial}{\partial y}$ and $\tilde{X}_2 = X_2 \frac{\partial}{\partial y}$, where X_1, X_2 are functions of (x,t) only.

Hence $[\tilde{X}_2, \tilde{X}_1] = 0$, and $[\tilde{A},\tilde{A}_u] = [\tilde{A},\tilde{A}_u]_x = [\tilde{A},\tilde{A}_{ux}] = 0$

$(18)_c$ gives

$$(24)\qquad \tilde{B}^* = -24u^3 \frac{\partial}{\partial y} - 12u^2\tilde{X}_2 - 2u\,\tilde{X}_{2xx} + 8\tilde{X}_4$$

where $\tilde{X}_4 = X_4 \frac{\partial}{\partial y}$, X_4 function of (x,t,y). The coefficient 8 is taken in accordance with [2].

Equation $(18)_d$ gives

$(25)_a$ $\quad 2X_{4y} - X_{2x} = 0$

$(25)_b$ $\quad 8X_2X_{4y} - X_{2xxx} - X_{2t} = 0$

$(25)_c$ $\quad 8X_1X_{4y} + 4X_{4x} - X_{1t} = 0$

$(25)_a \Rightarrow X_{4y} = \frac{1}{2} X_{2x}$, thus $X_4 = \frac{y}{2} X_{2x} + X_4^*$, where X_4^* depends only on x,t. Putting in $(25)_c$ gives $X_{2xx} = 0$, so that

$$(26) \quad X_2(x,t) = f(t)x + g(t) \quad , \quad X_4(x,t,y) = \frac{y}{2} f(t) + X_4^*(x,t)$$

and we obtain conditions

$$(27) \quad \begin{cases} 4f^2 = f_t \\ 4fg = g_t \\ 4fX_1 + 4X^*_{4x} - X_{1t} = 0 \end{cases}$$

<u>Two different solutions are possible</u>

(i) $f \equiv 0$, $g \equiv$ constant

(ii) $f(t) = -\frac{1}{4(t-t_o)}$, $g(t) = \frac{x_o}{4(t-t_o)}$ $\Rightarrow X_2 = -\frac{x - x_o}{4(t-t_o)}$

We observe that, in the first case, we have locally $\tilde{X}_2 = \lambda \tilde{X}_3$; by analycity, such a property is true in the whole open set $\tilde{\Omega}$. Thus, <u>if at a point $\tilde{m}$ we have locally the situation (i), then the property $\tilde{X}_2 \equiv \lambda \tilde{X}_3$ is satisfied in $\tilde{\Omega}$, thus in the whole space $\tilde{M}$</u>.

<u>Case (i) : $\tilde{X}_2 \equiv \lambda \tilde{X}_3$</u>

Properties : $\tilde{X}_2 = \lambda \tilde{X}_3$; $\tilde{B}_p = -(2\lambda + 6u)\tilde{X}_3$

$\tilde{B}_z = 6z \tilde{X}_3$; $\tilde{B}_u = (-72u^2 - 24u\lambda - 6p)\tilde{X}_3$

are satisfied in $\tilde{\Omega}$ and, by analycity, in $\tilde{M}$.

Thus, <u>in a neighbourhood of every point $\tilde{m}$ of $\tilde{M}$</u>, we have local coordinates such that

$$(28) \quad \begin{cases} \tilde{A} = 2\tilde{X}_1 + (2u\lambda + 3u^2)\tilde{X}_3 \\ \tilde{B} = 8\tilde{X}_4 + (-2p\lambda - 6up + 3z^2 - 24u^3 - 12u^2\lambda)\tilde{X}_3. \end{cases}$$

According to the remark at the end of §III, it is possible to use local coordinates such that $\tilde{X}_1 = 0$. If $\tilde{X}_3 = X_3 \frac{\partial}{\partial y}$ and $\tilde{X}_4 = X_4 \frac{\partial}{\partial y}$, condition (16) gives

$$X_{4x} = 0$$

Now, using a change of coordinates $y' = \varphi(t,y)$ such that

$$\varphi_y X_4 + \varphi_t = 0 \tag{29}$$

we obtain <u>local coordinates such that $\widetilde{X}_1 = \widetilde{X}_4 = 0$</u>. Then, by (16), <u>$X_3$ is function of y only</u>.

Finally, we observe that <u>such coordinates are well defined up to a change of coordinates of the form $y' = \varphi(y)$</u>. Thus the differential equation

$$dy = 0$$

has an intrinsic signification and determines <u>a global analytic trivialisation</u> : Hence :

<u>PROPOSITION I.</u> If $\widetilde{X}_3 \neq 0$ and $\widetilde{X}_2 = \lambda\widetilde{X}_3$, there exist on $\widetilde{M}$ global coordinates such that the pseudopotential equation has the form :

$$\boxed{dy = X_3(y)[(2u\lambda + 3u^2)dx + (-2p\lambda - 6up + 3z^2 - 24u^3 - 12u^2\lambda)dt}$$

<u>Case (ii) : $\widetilde{X}_2 \neq \lambda\widetilde{X}_3$</u>

In this case, properties

$$\begin{aligned} 2(t-t_0)\widetilde{X}_2 &= -\tfrac{1}{2}(x - x_0)\widetilde{X}_3 \\ 2(t-t_0)\widetilde{B}_p &= ((x - x_0) - 12u(t-t_0))\widetilde{X}_3 \\ 2(t-t_0)\widetilde{B}_z &= (12z(t-t_0) - 1)\widetilde{X}_3 \\ 2(t-t_0)\widetilde{B}_u &= (12u(x-x_0) + (-12p-144u^2)(t-t_0))\widetilde{X}_3 \end{aligned}$$

are satisfied in $\widetilde{\Omega}$ and, by analyticity, in $\widetilde{M}$. Thus, <u>in a neighbourhood of every point $\widetilde{m} \in \widetilde{M}$</u>, we have local coordinates such that

$$\left\{ \begin{aligned} 2(t-t_0)\widetilde{A} &= 4(t-t_0)\widetilde{X}_1 + (-u(x-x_0) + 6u^2(t-t_0))\widetilde{X}_3 \\ 2(t-t_0)\widetilde{B} &= 16(t-t_0)\widetilde{X}_4 + ((p+6u^2)(x-x_0)+2(3z^2-6up-24u^3)(t-t_0)-z)\widetilde{X}_3 \end{aligned} \right. \tag{30}$$

As in the previous case, we can use local coordinates such that $\widetilde{X}_1 = 0$. If $\widetilde{X}_3 = X_3 \frac{\partial}{\partial y}$ and $\widetilde{X}_4 = X_4 \frac{\partial}{\partial y}$, condition (16) gives

$$X_{4x} = 0$$

and, by a change of coordinates $y' = \varphi(t,y)$ satisfying (29), we obtain $\widetilde{X}_1 = \widetilde{X}_4 = 0$. Then, by (16), we have

$$X_{3x} = 0$$

$$X_3 - (t-t_0)X_{3t} = 0 \quad \Rightarrow \quad X_3(t,y) = 2(t-t_0)X_3^*(y)$$

and, by the same argument that in the first case, we obtain

<u>PROPOSITION II.</u> If $\tilde{X}_3 \neq 0$ and $\tilde{X}_2 \neq \lambda \tilde{X}_3$, there exist on $\tilde{M}$ global coordinates such that the pseudopotential equation has the form :

$$dy = X_3^*(y)[(-u(x-x_0)+6u^2(t-t_0))dx+((p+6u^2)(x-x_0)+ +(6z^2-12up-48u^3)(t-t_0)-z)dt]$$

V - GLOBAL MODELS IF $\tilde{X}_3 \equiv 0$ and $\tilde{X}_2 \neq 0$

In this case, $\tilde{\Omega} = \{\tilde{m} \in \tilde{M}/\tilde{X}_{2\tilde{m}} \neq 0\}$ is an open dense set in $\tilde{M}$.

If $\tilde{m} \in \tilde{\Omega}$, <u>it is possible to change locally coordinates</u>, in an open neighbourhood $\tilde{U}$ of $\tilde{m}$, <u>in order to have</u> $\tilde{X}_2 = \frac{\partial}{\partial y}$. In this case

$$\tilde{A} = 2\tilde{X}_1 + 2u\frac{\partial}{\partial y} \quad ; \quad \tilde{A}_u = 2\frac{\partial}{\partial y} \quad ; \quad [\tilde{A},\tilde{A}_u] = -4X_{1y}\frac{\partial}{\partial y} \quad \text{where} \quad \tilde{X}_1 = X_1\frac{\partial}{\partial y}.$$

If $\tilde{B}^* = B^*\frac{\partial}{\partial y}$, $(18)_c$ gives

$$(31) \quad B^* = 4u^2X_{1yy} - 12u^2 + 4uX_{1yx} + 8u(X_1X_{1yy} - X_{1y}^2) + 8X_4$$

where X_4 is a function of (x,t,y).

Finally, $(18)_d$ gives

$$(32) \quad \begin{cases} X_{1yyy} = 0 \\ X_{1yyx} + 2X_{1y} - 2X_{1y}X_{1yy} = 0 \end{cases}$$

First relation implies

$$(33) \quad X_1 = \alpha y^2 + \beta y + \gamma$$

where α, β, γ are functions of (x,t)

Second relation gives

$$(34) \quad \begin{cases} \alpha(\alpha - \frac{1}{2}) = 0 \\ \alpha_x + 4\beta(\frac{1}{2} - \alpha) = 0 \end{cases}$$

<u>Two cases are possible</u> : (i) $\alpha = 0$, $\beta = 0$ in a neighbourhood of $\tilde{m}$

(ii) $\alpha = \frac{1}{2}$ in a neighbourhood of $\tilde{m}$

Case (i) : $\alpha = \beta = 0$

We have $X_{1y} = 0 \Rightarrow B^* = -12u^2 + 8X_4$

By $(18)_d$, $X_{4y} = 0$. Hence, in the domain of local coordinates, we have

$$\begin{cases} \tilde{A} = 2\tilde{X}_1 + 2u\dfrac{\partial}{\partial y} \\ \tilde{B} = 8\tilde{X}_4 + (-2p - 12u^2)\dfrac{\partial}{\partial y} \end{cases}$$

Now, in this case, properties $\tilde{X}_3 = 0$ $\quad \tilde{B}_p = -2\tilde{X}_2$

$\tilde{B}_z = 0 \quad \tilde{B}_u = -12u\,\tilde{X}_2$

are true in a neighbourhood of $\tilde{m}$, and, by analycity, <u>in the whole manifold $\tilde{M}$</u>.

Thus, <u>in a neighbourhood of every point $\tilde{m}' \in \tilde{M}$</u>, we have local coordinates such that

$$(35) \quad \begin{cases} \tilde{A} = 2\tilde{X}_1 + 2u\,\tilde{X}_2 \\ \tilde{B} = 8\tilde{X}_4 + (-2p - 12u^2)\tilde{X}_2 \end{cases}$$

According to the remark at the end of §III, it is possible to use local coordinates such that $\tilde{X}_1 = 0$. Then, (16) gives $\tilde{X}_{4x} = 0$, and by a change of coordinates of the form $y' = \varphi(t,y)$ satisfying (29), <u>we obtain $\tilde{X}_1 = \tilde{X}_4 = 0$</u>. By (16), we have, if $\tilde{X}_2 = X_2\dfrac{\partial}{\partial y}$

$X_{2x} = X_{2t} = 0 \Rightarrow X_2$ is function of y only.

By the same argument that in the previous cases, we obtain

<u>PROPOSITION III</u>. If $\tilde{X}_3 \equiv 0$, $\tilde{X}_2 \neq 0$, $\alpha = \beta = 0$ at a point $\tilde{m}$, there exist on $\tilde{M}$ global coordinates such that **the pseudopotential equation has the form:**

$$\boxed{dy = X_2(y)[2udx + (-2p - 12u^2)dt]}$$

<u>Case (ii) : $\alpha = \frac{1}{2}$</u>

We have in this case, in the domain of local coordinates,

$A = y^2 + 2\beta y + 2\gamma + 2u$

and, by a change of coordinates of the form $y' = y + \beta$, we obtain a reduced expression in which $\beta = 0$:

$$(36) \quad \begin{cases} A = y^2 + 2\gamma + 2u \\ B = -2p - 4yz - 8u^2 - 4uy^2 + 8u\gamma + 8X_4 \end{cases}$$

Relation $(18)_d$ gives

$$\begin{cases} X_4 = \gamma y^2 - \dfrac{\gamma_x}{2} y + X_4^* \quad \text{where } X_4^* \text{ is function of } (x,t) \text{ only} \\ \gamma_x = 0 \\ 32\gamma^2 - 16X_4^* - 4\gamma_{xx} = 0 \\ 8\gamma\gamma_x + 8X_{4x}^* - 2\gamma_t = 0 \end{cases}$$

Thus, γ is a constant : $\gamma = \mu \in \mathbb{R}$, and we obtain the [local] reduced form

$$(37)\quad \begin{cases} A = y^2 + 2\mu + 2u \\ B = -2p - 4yz - 8u^2 - 4uy^2 + 8u\mu + 8\mu y^2 + 16\mu^2 \end{cases}$$

Now, if we define the [intrinsic] vector fields $\tilde{X}_5$, $\tilde{X}_6$ by

$$\tilde{X}_5 = y\frac{\partial}{\partial y} = -\frac{1}{4}\tilde{B}_z \quad \text{and} \quad \tilde{X}_6 = y^2\frac{\partial}{\partial y} = (2\mu - 4u)\tilde{X}_2 - \frac{1}{4}\tilde{B}_u$$

properties

$$(38)\qquad [\tilde{X}_2,\tilde{X}_5] = \tilde{X}_2 \quad ; \quad [\tilde{X}_2, \tilde{X}_6] = 2\tilde{X}_5 \ ; \ [\tilde{X}_5,\tilde{X}_6] = \tilde{X}_6$$

are true in a neighbourhood of $\tilde{m}$ and, by analyticy, in the whole manifold $\tilde{M}$.

Thus, in a neighbourhood of every point $\tilde{m}' \in \tilde{M}$, we have local coordinates such that we have (38) and

$$(39)\quad \begin{cases} \tilde{A} = 2\tilde{X}_1 + 2u\tilde{X}_2 \\ \tilde{B} = 8\tilde{X}_4 + (-2p - 8u^2 + 8u\mu)\tilde{X}_2 - 4z\tilde{X}_5 + (-4u)\tilde{X}_6 \end{cases}$$

Now, we define $\tilde{X}_1'$ and $\tilde{X}_4'$ by

$$(40)\quad \begin{cases} 2\tilde{X}_1 = 2\tilde{X}_1' + \tilde{X}_6 + 2\mu\tilde{X}_2 \\ 8\tilde{X}_4 = 8\tilde{X}_4' + 8\mu\tilde{X}_6 + 16\mu^2\tilde{X}_2 \end{cases}$$

in order to obtain a local expression like (37) :

$$(41)\quad \begin{cases} \tilde{A} = 2\tilde{X}_1' + \tilde{X}_6 + (2\mu + 2u)\tilde{X}_2 \\ \tilde{B} = 8\tilde{X}_4' + (-2p - 8u^2 + 8u\mu + 16\mu^2)\tilde{X}_2 - 4z\tilde{X}_5 + (8\mu - 4u)\tilde{X}_6 \end{cases}$$

According to the method indicated at the end of §III, it is possible to change coordinates in a neighbourhood $\tilde{U}'$ of $\tilde{m}'$ in order to obtain $\tilde{X}_1' = 0$. New coordinates are well defined up to change of the form $y' = \varphi(t,y)$. Near every point of $\tilde{U}' \cap \tilde{\Omega}$, there exist such coordinates with the additional property $\tilde{X}_4' = 0$. This fact implies that, if $\tilde{X}_1' = 0$ in $\tilde{U}'$, then $\tilde{X}_{4x}'$ is zero in $\tilde{U}' \cap \tilde{\Omega}$, thus $\tilde{X}_{4x}' = 0$ in $\tilde{U}'$. Now, by a change $y' = \varphi(y,t)$ such that

$$X_4'\varphi_y + \varphi_t = 0$$

we obtain, in a neighbourhood of $\tilde{m}'$, coordinates such that $\tilde{X}_4' = \tilde{X}_1' = 0$

As in the previous cases, we observe that such coordinates are defined up to a change of the form $y' = \varphi(y)$. Thus, differential equation $dy = 0$ determines a global analytic trivialisation. In the corresponding global coordinates (x,t,u,z,p,y) we have the reduced form

$$(42)\quad \begin{cases} \tilde{A} = (2\mu + 2u)\tilde{X}_2 + \tilde{X}_6 \\ \tilde{B} = (-2p - 8u^2 + 8u\mu + 16\mu^2)\tilde{X}_2 - 4z\tilde{X}_5 + (8\mu - 4u)\tilde{X}_6 \end{cases}$$

where $\tilde{X}_2$, $\tilde{X}_5$, $\tilde{X}_6$ satisfy (38) and $\mu \in \mathbb{R}$.

Moreover, at each point of $\tilde{\Omega}$, there exists a local change of coordinates $y' = \varphi(y)$ such that we obtain the reduced form (37). This implies that, if $\tilde{X}_i = X_i \frac{\partial}{\partial y}$, $i = 2,5,6$, we have in $\tilde{\Omega}$ [thus in $\tilde{M}$]

$$X_{2x} = X_{2t} = X_{5x} = X_{5t} = X_{6x} = X_{6t} = 0$$

Finally, we obtain :

PROPOSITION IV. If $\tilde{X}_3 \equiv 0$, $\tilde{X}_2 \neq 0$, $\alpha = \frac{1}{2}$ at a point $\tilde{m}$, there exist on $\tilde{M}$ global coordinates such that the pseudopotential equation has the form :

$$\boxed{\begin{aligned} dy = X_2(y)[(2\mu+2u)dx+(-2p-8u^2+8u\mu+16\mu^2)dt]+X_5(y)[-4zdt] \\ +X_6(y)[dx+(8\mu-4u)dt] \end{aligned}}$$

where the vector fields $\tilde{X}_i = X_i(y)\frac{\partial}{\partial y}$, $i = 2,5,6$, satisfy

$$[\tilde{X}_2,\tilde{X}_5] = \tilde{X}_2 \ ; \ [\tilde{X}_2,\tilde{X}_6] = 2\tilde{X}_5 \ ; \ [\tilde{X}_5,\tilde{X}_6] = \tilde{X}_6$$

This pseudopotential equation is essentially equivalent to the model discovered by E-W in [2].

VI - Main result and final observations.

In order to complete classification, we have to study the case $\tilde{X}_2 \equiv \tilde{X}_3 \equiv 0$. From (16) we obtain in this case that A, B are functions of (x,t,y) only. By the argument at the end of §III, we can use local coordinates such that $A = 0$. Then $B_x = 0$ and, by a change $y' = \varphi(t,y)$ such that $\varphi_y B + \varphi_t = 0$ we obtain $A = B = 0$. Hence $\mathcal{H}$ is an integrable connection. It defines global trivialisations of $(\tilde{M},\tilde{\pi},M)$ such that <u>the pseudopotential equation has the reduced form $dy = 0$</u>.

Now, we observe that, in propositions I, II, III, if $X_3(y) = 1$ or $X_3^*(y) = 1$ or $X_2(y) = 1$, the pseudopotential equation is <u>a potential equation</u> in the sense that

$$dA \wedge dx + dB \wedge dt = 0 \mod \tilde{\pi}^*\alpha,\ \tilde{\pi}^*\beta,\ \tilde{\pi}^*\gamma$$

If X_3, X_3^*, X_2 are non singular functions, by a change $y' = \varphi(y)$, we obtain a potential equation. These cases will be referred to as <u>potential cases</u>. Moreover, it is possible to put these three models together [including the trivial model $dy = 0$]

by introducing arbitrary contants. By this way, we obtain the following formulation of our main result :

THEOREM. Let $(\widetilde{M},\widetilde{\pi},M)$ be a locally trivial fiber bundle with R as typical fiber. If $\mathcal{H}$ is a Cartan-Ehresmann connection on $(\widetilde{M},\widetilde{\pi},M)$ adapted to the EDS (1), then there exist global coordinates (x,t,u,z,p,y) such that $\mathcal{H}$ is defined by one of the following pseudopotential equations :

(i) potential case :

$$dy = X(y)[[\lambda(-ux+6u^2t)+2u\mu+3u^2\nu]dx + [\lambda(px+6u^2x+6z^2t-12upt -48u^3-z)+\mu(-2p-12u^2)+\nu(-6up+3z^2-24u^3)]dt]$$

where $\lambda,\mu,\nu \in \mathbb{R}$ and X is an arbitrary function on $\mathbb{R}$.

(ii) pseudopotential case :

$$dy = X_2(y)[(2\mu+2u)dx+(-2p-8u^2+8u\mu+16u^2)dt]+X_5(y)[-4zdt]+X_6(y)[dx+(8\mu-4u)dt]$$

where $\mu \in \mathbb{R}$, and the vector fields $\widetilde{X}_i = X_i(y)\frac{\partial}{\partial y}$, $i = 2,5,6$, on $\mathbb{R}$ satisfy $[\widetilde{X}_2,\widetilde{X}_5] = \widetilde{X}_2$; $[\widetilde{X}_2,\widetilde{X}_6] = 2\widetilde{X}_5$; $[\widetilde{X}_5,\widetilde{X}_6] = \widetilde{X}_6$.

REMARKS 1 - Using the differentials of the right-hand members in the previous potential equations, we obtain closed 2-forms in the ideal $\mathcal{J}$ of (1) By this way, we obtain on M a new EDS :

$$(43)\quad \begin{cases} \alpha_1 \equiv -du \wedge dx + dp \wedge dt + 12udu \wedge dt = 0 \\ \beta_1 \equiv u\alpha_1 + pdu \wedge dt - zdz \wedge dt = 0 \\ \gamma_1 \equiv x\alpha_1 - 12t\beta_1 + pdx \wedge dt - dz \wedge dt = 0 \end{cases}$$

All the left-hand members in (43) are closed. Moreover, if $\mathcal{J}_1$ is the ideal of differential forms in M associated to (43), we have

$$\beta \in \mathcal{J}_1 \quad , \quad p\alpha \in \mathcal{J}_1 \quad , \quad \gamma + 12u\alpha \in \mathcal{J}_1$$

If S_1 is an integral manifold of (43), either S_1 is an integral manifold of (1), or we have on S_1

$$p = 0 \quad , \quad dz \wedge dt = 0 \quad , \quad du \wedge dx = 12u\, du \wedge dt$$

If $s_1 : \mathbb{R}^2 \to M$ is a section of π such that

$$s_1(x,t) = (x,t,u(x,t),z(x,t),p(x,t))$$

then, if $S_1 = s_1(\mathbb{R}^2)$ is an integral manifold of (43), either $u(x,t)$ is a solution of the Kd V-equation with $z = u_x$ and $p = u_{xx}$, or $u(x,t)$ is a solution of

$$(44)\qquad 12uu_x + u_t = 0$$

Hence, (43) gives another representation by an EDS of the KdV-equation, which introduces solutions of (44) as "parasites".

2 - It is an open question to ask if the previous result is true in the C^{∞}-case.

3 - It would be interesting to extend these arguments to other evolution equations in order to obtain all the possible simple analytic pseudopotentials, up to global equivalence.

REFERENCES

[1] E. CARTAN "Les systèmes différentiels extérieurs et leurs applications geometriques" Hermann, Paris (1945)

[2] H.D. WAHLQUIST - F.B. ESTABROOK "Prolongation structures of non-linear evolution equations" J. Math. Phys. 16, 1-7 (1975).

[3] R. HERMANN "Geometric theory of non-linear evolution equations, Bäcklund transformations and solitons" A-B, Vol XII, XIV Interscience Math., Brookline (1976-77).

Mathématiques
Université des Sciences et Techniques
du Languedoc
Place E. Bataillon, 34060 Montpellier
FRANCE

Vol. 817: L. Gerritzen, M. van der Put, Schottky Groups and Mumford Curves. VIII, 317 pages. 1980.

Vol. 818: S. Montgomery, Fixed Rings of Finite Automorphism Groups of Associative Rings. VII, 126 pages. 1980.

Vol. 819: Global Theory of Dynamical Systems. Proceedings, 1979. Edited by Z. Nitecki and C. Robinson. IX, 499 pages. 1980.

Vol. 820: W. Abikoff, The Real Analytic Theory of Teichmüller Space. VII, 144 pages. 1980.

Vol. 821: Statistique non Paramétrique Asymptotique. Proceedings, 1979. Edited by J.-P. Raoult. VII, 175 pages. 1980.

Vol. 822: Séminaire Pierre Lelong-Henri Skoda, (Analyse) Années 1978/79. Proceedings. Edited by P. Lelong et H. Skoda. VIII, 356 pages, 1980.

Vol. 823: J. Král, Integral Operators in Potential Theory. III, 171 pages. 1980.

Vol. 824: D. Frank Hsu, Cyclic Neofields and Combinatorial Designs. VI, 230 pages. 1980.

Vol. 825: Ring Theory, Antwerp 1980. Proceedings. Edited by F. van Oystaeyen. VII, 209 pages. 1980.

Vol. 826: Ph. G. Ciarlet et P. Rabier, Les Equations de von Kármán. VI, 181 pages. 1980.

Vol. 827: Ordinary and Partial Differential Equations. Proceedings, 1978. Edited by W. N. Everitt. XVI, 271 pages. 1980.

Vol. 828: Probability Theory on Vector Spaces II. Proceedings, 1979. Edited by A. Weron. XIII, 324 pages. 1980.

Vol. 829: Combinatorial Mathematics VII. Proceedings, 1979. Edited by R. W. Robinson et al.. X, 256 pages. 1980.

Vol. 830: J. A. Green, Polynomial Representations of GL_n. VI, 118 pages. 1980.

Vol. 831: Representation Theory I. Proceedings, 1979. Edited by V. Dlab and P. Gabriel. XIV, 373 pages. 1980.

Vol. 832: Representation Theory II. Proceedings, 1979. Edited by V. Dlab and P. Gabriel. XIV, 673 pages. 1980.

Vol. 833: Th. Jeulin, Semi-Martingales et Grossissement d'une Filtration. IX, 142 Seiten. 1980.

Vol. 834: Model Theory of Algebra and Arithmetic. Proceedings, 1979. Edited by L. Pacholski, J. Wierzejewski, and A. J. Wilkie. VI, 410 pages. 1980.

Vol. 835: H. Zieschang, E. Vogt and H.-D. Coldewey, Surfaces and Planar Discontinuous Groups. X, 334 pages. 1980.

Vol. 836: Differential Geometrical Methods in Mathematical Physics. Proceedings, 1979. Edited by P. L. García, A. Pérez-Rendón, and J. M. Souriau. XII, 538 pages. 1980.

Vol. 837: J. Meixner, F. W. Schäfke and G. Wolf, Mathieu Functions and Spheroidal Functions and their Mathematical Foundations Further Studies. VII, 126 pages. 1980.

Vol. 838: Global Differential Geometry and Global Analysis. Proceedings 1979. Edited by D. Ferus et al. XI, 299 pages. 1981.

Vol. 839: Cabal Seminar 77 – 79. Proceedings. Edited by A. S. Kechris, D. A. Martin and Y. N. Moschovakis. V, 274 pages. 1981.

Vol. 840: D. Henry, Geometric Theory of Semilinear Parabolic Equations. IV, 348 pages. 1981.

Vol. 841: A. Haraux, Nonlinear Evolution Equations- Global Behaviour of Solutions. XII, 313 pages. 1981.

Vol. 842: Séminaire Bourbaki vol. 1979/80. Exposés 543–560. IV, 317 pages. 1981.

Vol. 843: Functional Analysis, Holomorphy, and Approximation Theory. Proceedings. Edited by S. Machado. VI, 636 pages. 1981.

Vol. 844: Groupe de Brauer. Proceedings. Edited by M. Kervaire and M. Ojanguren. VII, 274 pages. 1981.

Vol. 845: A. Tannenbaum, Invariance and System Theory: Algebraic and Geometric Aspects. X, 161 pages. 1981.

Vol. 846: Ordinary and Partial Differential Equations, Proceedings. Edited by W. N. Everitt and B. D. Sleeman. XIV, 384 pages. 1981.

Vol. 847: U. Koschorke, Vector Fields and Other Vector Bundle Morphisms – A Singularity Approach. IV, 304 pages. 1981.

Vol. 848: Algebra, Carbondale 1980. Proceedings. Ed. by R. K. Amayo. VI, 298 pages. 1981.

Vol. 849: P. Major, Multiple Wiener-Itô Integrals. VII, 127 pages. 1981.

Vol. 850: Séminaire de Probabilités XV. 1979/80. Avec table générale des exposés de 1966/67 à 1978/79. Edited by J. Azéma and M. Yor. IV, 704 pages. 1981.

Vol. 851: Stochastic Integrals. Proceedings, 1980. Edited by D. Williams. IX, 540 pages. 1981.

Vol. 852: L. Schwartz, Geometry and Probability in Banach Spaces. X, 101 pages. 1981.

Vol. 853: N. Boboc, G. Bucur, A. Cornea, Order and Convexity in Potential Theory: H-Cones. IV, 286 pages. 1981.

Vol. 854: Algebraic K-Theory. Evanston 1980. Proceedings. Edited by E. M. Friedlander and M. R. Stein. V, 517 pages. 1981.

Vol. 855: Semigroups. Proceedings 1978. Edited by H. Jürgensen, M. Petrich and H. J. Weinert. V, 221 pages. 1981.

Vol. 856: R. Lascar, Propagation des Singularités des Solutions d'Equations Pseudo-Différentielles à Caractéristiques de Multiplicités Variables. VIII, 237 pages. 1981.

Vol. 857: M. Miyanishi. Non-complete Algebraic Surfaces. XVIII, 244 pages. 1981.

Vol. 858: E. A. Coddington, H. S. V. de Snoo: Regular Boundary Value Problems Associated with Pairs of Ordinary Differential Expressions. V, 225 pages. 1981.

Vol. 859: Logic Year 1979–80. Proceedings. Edited by M. Lerman, J. Schmerl and R. Soare. VIII, 326 pages. 1981.

Vol. 860: Probability in Banach Spaces III. Proceedings, 1980. Edited by A. Beck. VI, 329 pages. 1981.

Vol. 861: Analytical Methods in Probability Theory. Proceedings 1980. Edited by D. Dugué, E. Lukacs, V. K. Rohatgi. X, 183 pages. 1981.

Vol. 862: Algebraic Geometry. Proceedings 1980. Edited by A. Libgober and P. Wagreich. V, 281 pages. 1981.

Vol. 863: Processus Aléatoires à Deux Indices. Proceedings, 1980. Edited by H. Korezlioglu, G. Mazziotto and J. Szpirglas. V, 274 pages. 1981.

Vol. 864: Complex Analysis and Spectral Theory. Proceedings, 1979/80. Edited by V. P. Havin and N. K. Nikol'skii, VI, 480 pages. 1981.

Vol. 865: R. W. Bruggeman, Fourier Coefficients of Automorphic Forms. III, 201 pages. 1981.

Vol. 866: J.-M. Bismut, Mécanique Aléatoire. XVI, 563 pages. 1981.

Vol. 867: Séminaire d'Algèbre Paul Dubreil et Marie-Paule Malliavin. Proceedings, 1980. Edited by M.-P. Malliavin. V, 476 pages. 1981.

Vol. 868: Surfaces Algébriques. Proceedings 1976–78. Edited by J. Giraud, L. Illusie et M. Raynaud. V, 314 pages. 1981.

Vol. 869: A. V. Zelevinsky, Representations of Finite Classical Groups. IV, 184 pages. 1981.

Vol. 870: Shape Theory and Geometric Topology. Proceedings, 1981. Edited by S. Mardešić and J. Segal. V, 265 pages. 1981.

Vol. 871: Continuous Lattices. Proceedings, 1979. Edited by B. Banaschewski and R.-E. Hoffmann. X, 413 pages. 1981.

Vol. 872: Set Theory and Model Theory. Proceedings, 1979. Edited by R. B. Jensen and A. Prestel. V, 174 pages. 1981.